ALTERNATIVE FUELS, TECHNICAL AND ENVIRONMENTAL CONDITIONS

Edited by **Krzysztof Biernat**

Alternative Fuels, Technical and Environmental Conditions
http://dx.doi.org/10.5772/60628
Edited by Krzysztof Biernat

Contributors
Maciej Mikulski, Sławomir Wierzbicki, Marta Ambrosewicz-Walacik, Kamil Duda, Andrzej Piętak, Maciej Paczuski, Andrzej Pankowski, Ryszard Puławski, Kamil Kurpiel, Marcin Przedlacki, Marcin Marchwiany, Ph.D. Anna Matuszewska, Miloslaw Kozak, Jerzy Merkisz, Pawel Fuc, Piotr Lijewski, Michal Dobrzynski

CBS, Edition **2017**

Published by InTech
Janeza Trdine 9, 51000 Rijeka, Croatia

Publishing Process Manager Iva Simcic

Technical Editor InTech DTP team

Cover Designer InTech Design Team

Additional hard copies can be obtained from orders@intechopen.com

Alternative Fuels, Technical and Environmental Conditions, Edited by Krzysztof Biernat
p. cm.
ISBN 978-953-51-2269-2

INTECH

open science | open minds

Contents

Preface

Fuels for internal combustion engines must meet certain requirements, directly related to the processes of evaporation and combustion of fuel and indirectly to ensure an adequate performance, fuel consumption and environmental protection, as well as ensuring the rational exploitation in general.

Development and modifications of modern engines drives therefore require the appropriate fuels to these drives, so that the above requirements have been met. Those limits are specified by designers and manufacturers of vehicles, while ensuring the durability and reliability of engines fuelled with appropriate fuel.

Researches on processes for obtaining substitutes of previously used fuels, defined as "alternative fuel" for motor petrol, diesel fuels, and even aviation fuels, are carried out intensively around the world. The alternative fuels include biofuels. The alternative fuel that is already widely used is LPG and recently compressed natural gas (CNG) and liquid nitrogen gas (LNG) are used. An alternative gaseous fuel for diesel engines is dimethyl ether (DME) or BioDME (already produced in the world on a pilot scale), wherein the DME can be produced from coal and BioDME from lignocellulosic biomass or even waste. It is proposed to purification of biogas to obtain clean biomethane and compressing it to a gaseous fuel with similar quality characteristics as CNG. In terms of liquid fuels, are conducted and implemented the production technologies of second-generation bioethanol from lignocellulose (waste biomass or specific energy crops) or from waste substances. Research also includes production of biobuthanol from fermentation processes and dimethylfuran (DMF) obtained from cellulose and starch; there are also contemplated the solar fuels, obtained in processes of thermal decomposition of biomass or wastes. Processes of converting biomass to liquid fuels are known as "biomass to liquid" (BtL) processes and derived from waste—"waste to liquid" (Wtl). Also contemplated are alternative fuels, which are suitably composed of mixtures of synthetic hydrocarbons, obtained from a various materials—the so-called xtL process. Perspective raw material can be synthesis gas obtained from a wide variety of thermal processes and thermocatalytic waste substances, including biomass and energy crops, and also what is new, coming from the synthesis of water vapor and carbon dioxide. From the synthesis gas can be obtained various kinds of hydrocarbons, compositions of which will correspond to the composition of petrol, diesel oil and aviation fuels. Research is conducted on the possible fermentation of synthesis gas, leading to the preparation of ethanol and hydrocarbons from C_2 to C_5. In the perspective there is hydrogen or biohydrogen, also obtained from synthesis gas or other biomass conversion processes, as universal energy carrier.

In terms of the most promising raw materials for obtaining of biofuels as future alternative fuels, due to the requirements for reducing CO_2 emissions, there are preferred cultivations of algae, *Camelina, Jatropha* and halophytes. For increasing the amount of usable biomass raw materials, there are also being developed such technologies as without sunlight (dark) photosynthesis and marine membrane systems for the algae production and technologies for bio-methanol as raw material.

As it is clear from the foregoing considerations, there is a noticeable trend in technology development and research of future alternative fuels with a greater range of raw materials than biofuels, which was the guiding thought in terms of the need to develop a monograph describing these issues.

The publishing intention was therefore to develop a monograph that contains an overview of technologies of different alternative fuels, together with the presentation of current research results in the field of exploitation conditions for these fuels and the impact of operational processes on the environment. However, currently the main research is in the field of advanced biofuels, and therefore the content of the monograph has been limited to present the views and the results of current research on gas fuel type (LPG and CNG), as well as operational and environmental conditions on the use of RME as a alternative fuel, carried out in Polish research centers. The monograph also contains a review of the possibilities for the production of selected alternative fuels derived from microorganisms, mainly microalgae as a feedstock. Therefore, monograph comprises six monothematic chapters.

The first chapter is devoted to reflection on the quality and exploitation conditions of LPG fuel in spark-ignition engines. This chapter presents the results of the authors' research, especially in terms of opportunities to reduce toxic exhaust emissions.

In the second chapter, the authors present a description of the problem and the results of their research in the field of environmental aspects of use of CNG in public transport. Research results, presented in this section, are derived from the processes of actual use of diesel engines in city buses.

The third chapter describes the results of studies on the opportunities to reduce toxic exhaust emissions in diesel engines of dual-fuel type, in the operating conditions. Studies conducted and presented by the authors include the possible use of gaseous fuels in the engines, while the selection of appropriate parameters determining the process of ignition of fuel dose.

The fourth chapter describes the problems and the results of research on exploitation conditions of diesel engines supplied by RME fuel in public transport, especially in the area of impact on the environment.

The fifth chapter contains a discussion on the impact of derivatives of oxygenated compounds of hydrocarbons on the fuel combustion processes in diesel engines in the field of changes in toxic exhaust emissions.

The sixth chapter is a comprehensive review on the material range of possibilities for obtaining various alternative fuels falls within the area of biofuels using microalgae as raw materials directly and also indirectly. The problems described in this section are located in the directions of development of the future technologies, advanced liquid fuels for transport.

In conclusion, on behalf of myself and the authors of this monograph, I would like to thank Mrs. Iva Simcic from InTech Publishing, for very efficient and effective completion of the publishing process as well as patience and understanding.

I dedicate this monograph to my grandson, Natan, with the hope that his generation will have a sufficient amount of high-quality alternative fuels, fully environmentally friendly.

Prof. Krzysztof Biernat
Automotive Industry Institute
Institute for Ecology and Bioethics
of Cardinal Stefan Wyszynski University
Warsaw, Poland

Chapter 1

Environmental Aspects of the Use of CNG in Public Urban Transport

Jerzy Merkisz, Michał Dobrzyński, Miłosław Kozak, Piotr Lijewski and Paweł Fuć

Additional information is available at the end of the chapter

http://dx.doi.org/10.5772/62978

Abstract

This chapter concerns the problem of assessing the exhaust emission from the engines of city transport buses fuelled by CNG. It presents a comparative analysis of toxic exhaust emissions of CO, HC, NO_x and PM, from urban buses powered by diesel and CNG. The measurements were carried out over the SORT standardised cycles as well as during a real drive condition on a city bus route. The research revealed that CNG bus generates significantly lower NO_x emission, whereas its CO and HC emissions are higher. Taking into account low PM emissions, CNG buses should be regarded as eco-friendly means of public transport.

Keywords: Bus, CNG, exhaust emissions, PEMS, RDE

1. Introduction

The idea of fueling engines with gaseous fuels is as old as the history of the piston engine itself. The first combustion engine in the world — the one built in 1860 by Etienne Lenoir — was fueled with light gas, similarly to the first four-stroke engine built by Nicolaus Otto in 1876. The construction of a gasoline engine by Carl Benz practically eliminated the use of gaseous fuels for almost a hundred years. The interest in gaseous fuels for spark ignition engines returned last century, in the 1970s, in the times of energy crisis and this interest has continued to date.

Gaseous fuels belong to a group of most important alternative fuels for spark ignition engines (Figure 1) while they are almost neglected for diesel engines. Of all gaseous fuels, gaseous hydrocarbons are most frequently used such as: methane (*natural gas*) and LPG (*liquefied petroleum gas*).

In Poland a very popular gaseous fuel for motor vehicles is LPG. The number of vehicles adapted for this fuel exceeded 2 million already a few years ago. In this respect Poland places in the top 3 along with South Korea and Turkey. In terms of the number of LPG filling stations, Poland places first worldwide *ex equo* with Turkey [1]. Natural gas on the other hand is not a very popular engine fuel in Poland. The filling station infrastructure is still poorly developed. The only sector of road transport where the use of natural gas as a fuel increases is municipal transit (Rzeszów, Przemyśl and Kraków city bus operators) [2].

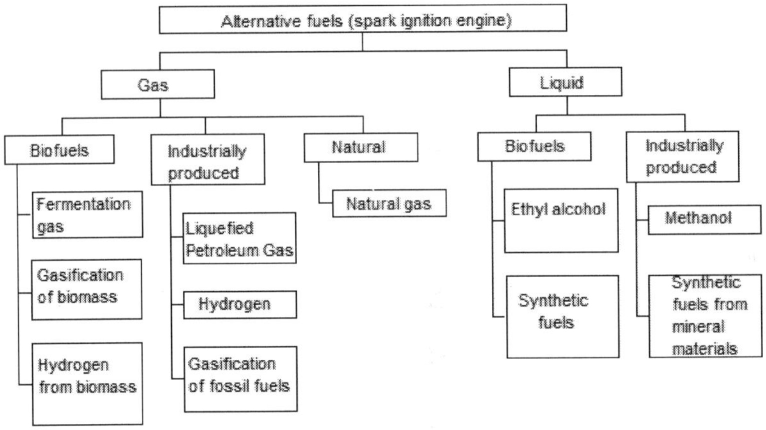

Figure 1. Gaseous fuels compared to other alternative spark ignition engine fuels.

The attractiveness of gaseous fuels is mainly attributed to their low price. These fuels are cheaper than conventional liquid fuels practically in the whole world. In Europe LPG is, on average, twice cheaper than gasoline (Figure 2). This is also the case for natural gas [3]. This results not only from the lower costs of production but also from the fiscal policy of many countries.

Fuels used to power combustion engines must have specific properties. This is also the case for gaseous fuels. The most widely used gaseous engine fuels — hydrocarbon fuels (LPG, natural gas) — owing to their properties (Table 1), are suitable for application in spark ignition engines. LPG is obtained as a product of crude oil processing. Natural gas is a standalone mineral fuel but is frequently found along with crude oil. The main component of natural gas is methane. It also contains other light hydrocarbons in smaller amounts, particularly ethane, propane, butane, pentane and hexane, as well as nitrogen and helium.

Among the main features of hydrocarbon fuels one should list the following [5]:

- High octane number

http://dx.doi.org/10.5772/62978

- Hydrocarbon fuels easily mix with air, which makes the mixture homogenous and its composition in the micro and macro scale is level
- The combustion is usually smokeless
- Hydrocarbon fuels generate a lower mass of exhaust gas compared to liquid fuels
- The air-gas mixtures have wide ranges of ignition, which enables combustion of leaner mixtures
- Gases have high ignition temperatures and, when used in diesel engines, require positive ignition

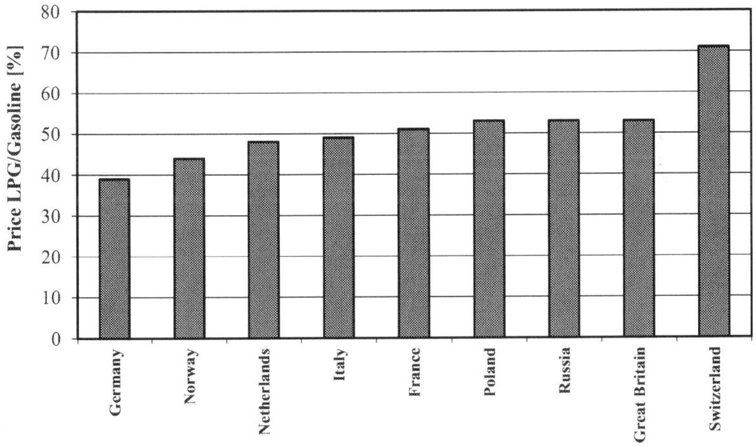

Figure 2. Price of 1 dm^3 of LPG as a percent of 95 gasoline fuel in different European states [4].

It is a common knowledge that vehicles powered with alternative fuels are environment friendly. Owing to the popularity of LPG- fueled engines, it is for these particular engines that we have most data available. These engines are usually fed with stoichiometric mixture and are fitted with three-way catalysts (TWC). The advantage of LPG in terms of low emission is high value of the H/C ratio and simple chemical composition. The exhaust emission level, however, heavily depends on the LPG system quality (the accuracy of fuel dosage, ignition timing, valve timing etc.) Accuracy that is comparable to that of gasoline fuel systems can only be achieved in the 4th and 5th generation LPG systems, particularly if the engine maps are optimized for LPG feed. In such a case, it is possible to obtain exhaust emissions of comparable or even lower values than for gasoline (Figure 3).

There are much less exhaust emission data for natural gas-fueled vehicles, particularly heavy-duty vehicles. This is partly attributed to rather low popularity of NG -fueled heavy-duty vehicles but also to much higher costs and greater complexity of the exhaust emission testing

Fuel	Density under ordinary conditions [kg/m³]	Fuel calorific value [MJ/kg]	Boiling temperature [°C]	Air demand [kg/kg]	Excess air coefficient λ up to the ignitability boundaries	Octane number MON (RON)
Methane	0.72	50	−161	17.2	0.7 – 2.1	110 (140)
Liquefied petroleum gas Propane-butane 50%/50%	2.35	46	−30	15.5	0.4 – 1.7	95 (100)
Natural gas[1)]	0.74	48	−161	17.0	0.7 – 2.1	100–110
Generator gas[1)]	1.02	5.5	- 253 – -192	1.20	–	105
Fermentation gas [1)] (biogas)	1.20	20	–	6.6	0.7 – 2.3	110
Hydrogen	0.089	120	−253	34	0.5 – 10.5	70
Gasoline 95	720 – 775	42.6	40 – 210	14.7	0.4 – 1.4	85 (95)
Methanol	797	19.9	64.4	6.5	0.3 – 2.0	95 (115)
Ethanol	793	26.7	78.3	9.0	0.3 – 2.1	94 (111)

[1)] Typical values have been given; gas composition and its parameters may vary widely.

Table 1. Properties of gaseous fuels compared to other spark ignition engine fuels [6]

in this group of vehicles. In general, we can state that NG -fueled heavy-duty vehicles are predominantly single fuel units based on either stoichiometric or lean mixture strategy.

Engines fed with a stoichiometric mixture ($\lambda = 1$) are fitted with three-way catalysts (TWC). Compared to conventional HDD engines, they are characterized by much lower emission of NO_x and PM but also by 15-20% higher energy consumption (under the conditions of the ESC test). The conversion rate of CO, HC and NO_x depends on the capability of keeping $\lambda = 1$. Due to a high temperature of the exhaust gas, premature process of deactivation of the catalytic converter and the oxygen sensor may occur leading to an increase in the exhaust emissions [8].

Engines fed with homogenous lean mixtures operate with the excess air coefficient in the range from $\lambda = 1.6$ (full load) to $\lambda = 1.2$ (idle) i.e. in the range ensuring low emission of CO and NO_x. The elimination of CO and HC is done with the oxicat converters, which, however, have a low conversion rate of methane. For this reason, the emission of THC from gas- fueled engines remains on a level similar to diesel- fueled engines, while the emission of the outstanding exhaust components is much lower. The consumption of gas is lower than in the earlier described stoichiometric scenario [8].

As mentioned in the beginning, in Poland the only sector of transport where the number o natural gas-fueled vehicles is growing dynamically is public transport. These are usually new CNG- fueled vehicles of very high emission categories (Euro V, EEV). The aim of this paper is to experimentally determine the exhaust emission parameters of this type of vehicles.

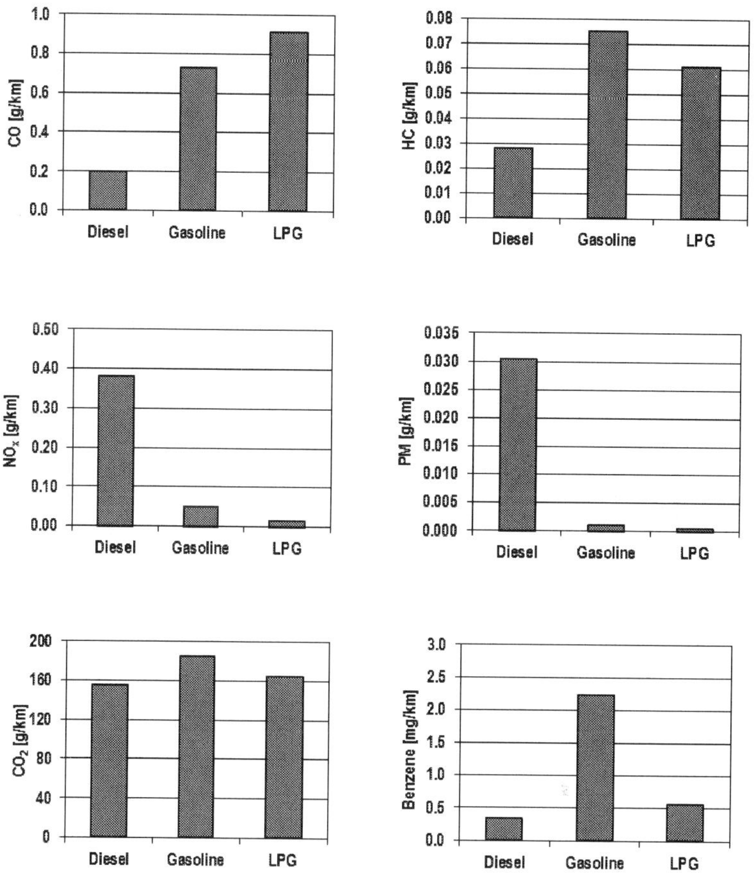

Figure 3. Average exhaust emissions in the NEDC test from modern passenger vehicles fueled with conventional fuels and LPG [7].

2. Overview of CNG vehicles

Given the availability of significant amounts of natural gas in the world, high level of operational safety and low emissions, CNG has the greatest share in the three mentioned alternative fuels. The CNG feed is realized from gas cylinders in which the gas is compressed to the pressure of approx. 20–25 MPa. The most frequently modified engines are spark ignition and diesel engines including their (single or dual) fuel systems. Dual fuel systems are most frequently used in passenger vehicles. Buses and other transport vehicles are usually fitted with single, gas fuel systems. Currently, an increasing number of carmakers have serially manufactured CNG-fueled vehicles in their portfolio. Natural gas is mainly used as bus fuel in short distance carriage of passengers. This results mainly from the economic factors. The cost of CNG in transport is much lower than that of diesel fuel. This chiefly results from the lack of sufficiently developed fueling stations. This is one of the most important issues blocking rapid development of CNG vehicles for carriage of goods and passengers. This problem extends across all European states. The number of CNG vehicles operated in Europe is growing, but it is still relatively low. Table 2 presents the data related to the number of CNG vehicles in selected European states.

Currently, most of the manufacturers of light duty vehicles (LDV) and buses have CNG-fueled vehicles in their portfolio. Works are continuing on the improvement of the designs of these vehicles and their engines. One of the most significant problems is fitting large CNG tanks in the vehicle. Despite the fact that the tanks must operate under high pressure, they must also be of high volume. For larger vehicles (buses, light duty trucks, heavy duty trucks) there is usually enough space to fit the tanks, but for light duty vehicles this remains a serious problem. The fitting of the tanks must not compromise the vehicle safety and its functionality. Examples of vehicle designs with the CNG tanks visible have been shown in Figure 4.

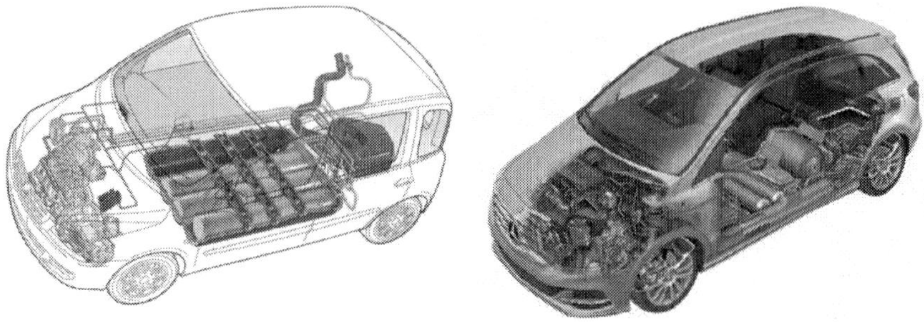

Figure 4. Examples of CNG -fueled vehicles.

Another factor influencing the common use of CNG as fuel is the possibility of quick refueling. Two types of refueling can be distinguished: quick (vehicle is fueled in 3–7 minutes) and slow (usually lasts up to several minutes for a passenger car). In order to quickly charge natural gas into the gas tank without high efficiency compressor, cascade storage is applied. Cascade storage is a tri-segment (low-, mid- and high pressure) system. The system includes 10–50 tanks of the capacity of 80–150 dm3 each for each segment. Between fuelings, the compressor charges the cascade system (to the pressure of 25–30 MPa) and during the refueling process, the gas is fed from the cascade storage and the compressor only maintains the required pressure. Slow refueling systems do not require cascade storage. The gas is charged directly to the tanks with the pressure slightly exceeding nominal. Tanks fitted in vehicles are filled within 6–8 h, which is a much cheaper solution for vehicle fleets or private users.

http://dx.doi.org/10.5772/62978

UE countries	LDV	Buses	Trucks
Austria	8 100	167	54
Belgium	1015	3	15
Bulgaria	61 000	280	40
Croatia	219	78	3
Cyprus	0	0	0
Czech	6650	512	81
Denmark	61	26	17
Estonia	300	30	10
Finland	1 600	45	20
France	10 050	2 400	1 100
Germany	95 708	1 735	176
Greece	280	618	102
Hungary	5 000	86	32
Ireland	3	0	0
Italy	880 000	2 300	3 000
Latvia	29	0	0
Lithuania	80	300	0
Luxembourg	230	39	1
Malta	0	0	0
Holland	6 498	686	386
Poland	3 050	400	50
Portuguese	45	354	86
Romania	0	0	0
Slovakia	1 100	261	65
Slovenia	29	24	5
Spain	905	1 609	1 322
Sweden	43 795	755	2 163
Great Britain	20	37	621
Summary	1 125 768	12 745	9 349

Table 2. Number of CNG -fueled vehicles in the EU member states divided into types of vehicles

3. Testing methods and equipment

The investigations were performed on two 18m city buses fitted with a CNG and a diesel engine respectively. The bus fitted with the CNG engine (8.9 dm^3) produced 238 kW of power. The vehicle was fitted with a three- way catalyst (TWC). The composite CNG cylinders of the capacity of 214 dm^3 each were mounted on the roof. The cylinders fitted on the roof increased the vehicle's height to 3 400 mm, which may constitute a serious downside in terms of road infrastructure. Basic technical data of the engines of the tested buses have been presented in Table 3. The view of the buses with the measurement equipment has been shown in Figure 4.

Parameter	**CNG -fueled vehicle A**	**Diesel- fueled vehicle B**
Engine type	Spark ignition	Compression ignition
Engine displacement	8.9 dm^3	9.2 dm^3
Number of cylinders	6	6
Compression ratio	12:1	17.5±0.5
Maximum power	239 kW @2000 rpm	231 kW @1900 rpm
Maximum torque	1356 Nm @1300 rpm	1275 Nm @1100–1710 rpm
Emission Technology	EEV	EEV
Aftertreatment	TWC	SCR/DPF
Length	18 000 mm	18 000 mm
Height	3 400 mm	3 050 mm
Vehicle weight	24 000 kg	24 000 kg

Table 3. Technical specifications of the tested city buses

In the on-road exhaust emissions tests of the city buses a PEMS portable exhaust emissions analyzer was used (Semtech DS by Sensors Inc.) capable of measuring and recording the following parameters (Figure 5):

- Concentration of CO and CO_2 (NDIR analyzer – Non-Dispersive Infrared), NO_x = NO + NO_2 (NDUV — Non-Dispersive Ultraviolet), HC (FID analyzer — Flame Ionization Detector), O_2 (electrochemical sensor);
- Thermodynamic parameters of the exhaust gas (mass flow, temperature, pressure) — a flow meter utilizing the Pitot tube;
- Ambient conditions — pressure, temperature, humidity;
- Position and speed of the vehicle — GPS system;
- Data pulled from the vehicle on-board diagnostic system (data transmission protocol — CAN SAE J1939/J2284.)

The diagram showing the location of the analyzer in the bus has been shown in Figure 6.

Figure 5. City buses prepared for the tests (left - CNG -fueled vehicle).

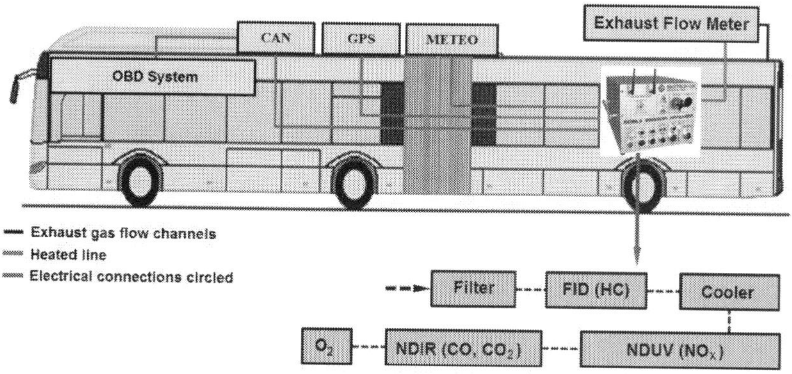

Figure 6. Diagram showing the location of the portable analyzer (Semtech DS) in the tested city buses.

The exhaust gas sample is taken from the flow meter and is then transported through a heated line (red line) maintaining the temperature of ~190°C. The temperature prevents possible HC condensation on the walls. Upon passing the filter, the sample reaches the FID analyzer, where the HC concentration is measured. When the sample is chilled to the temperature of 4°C it reaches the NDUV and NDIR analyzers respectively. These analyzers measure the concentrations of $NO_x = (NO + NO_2)$, CO and CO_2. Finally, a measurement of the O_2 concentration is performed with an electrochemical sensor. The control and monitoring of the Semtech DS device is realized with a portable computer connected to the main unit through a wireless connection. The device can communicate with a local area network; however, in the performed tests this method of communication was not applied.

The measurements were performed in the SORT tests and under actual conditions of operation (Figure 7), when the buses operated on regular city routes in Poznań. The selected test route reflected typical city bus operating conditions in large agglomerations. The selected city route has been classified by the bus operator as one of the most heavily loaded in terms of number of passengers and length (16.2 km). The test route had 43 bus stops. A varied configuration of the test route (main city road, residential area roads and downtown roads) ensures high variability of accelerations and share of road congestions, which enables the analysis of the exhaust emissions in a wide range of vehicle operating parameters.

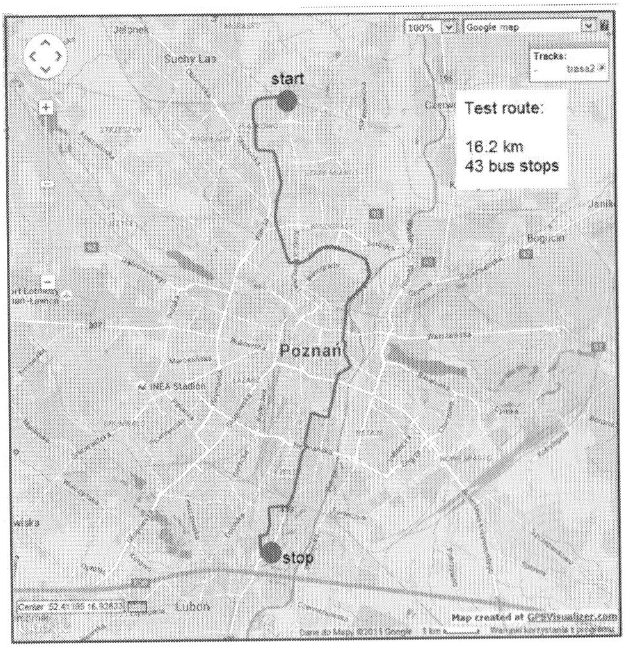

Figure 7. The RDE test route (created at gpsvisualizer.com).

The second stage of the investigations covered the SORT test measurements that are a universal and commonly recognized method of assessment of gas mileage as well as exhaust emissions. These tests are made of modules and reflect three types of traffic conditions – downtown roads, urban roads and extra urban roads (Figure 8; Table 4). The basic module of the SORT cycle is described with vehicle speed, route length and time. These parameters build the velocity profile characteristics of a given route, allowing for the bus stops, stops at intersections, driving off and cruise.

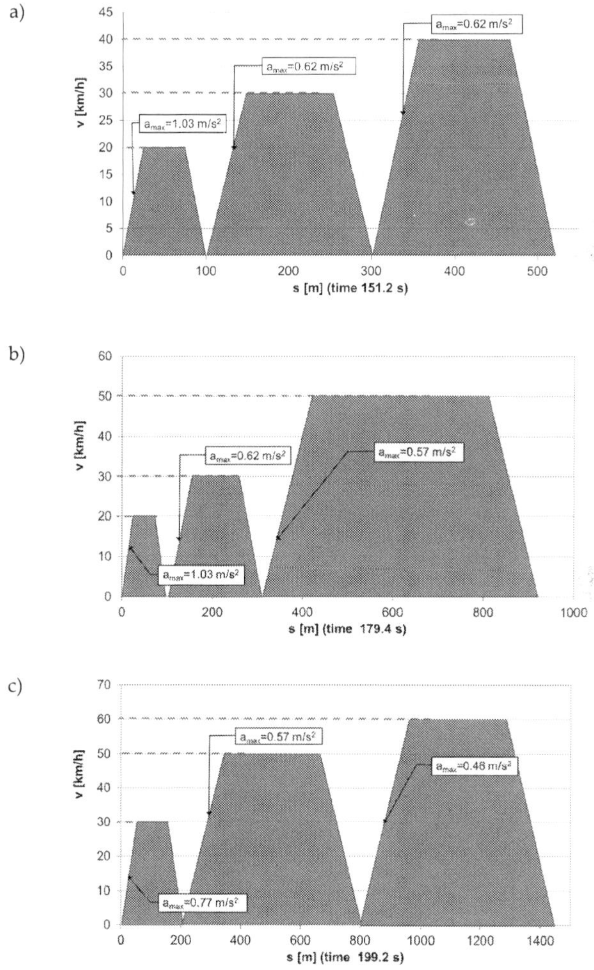

Figure 8. Velocity profiles of the SORT driving tests: a) SORT 1, b) SORT 2, c) SORT 3 [9].

	SORT 1	SORT 2	SORT 3
Average speed [km/h]	12.6	18.6	26.3
Share of stationary vehicle in the test [%]	39.7	33.4	20.1
Constant speed in profile 1 [km/h]/[m]	20/100	20/100	30/200
Acceleration in profile 1 [m/s²]	1.03	1.03	0.77
Constant speed in profile 2[km/h]/[m]	20/200	40/220	50/600
Acceleration in profile 2 [m/s²]	0.77	0.62	0.57
Constant speed in profile 3 [km/h]/[m]	40/220	50/600	60/650
Acceleration in profile 3 [m/s²]	0.62	0.57	0.46
Time of stoppage in each profile [s]	20/20/20	20/20/20	20/10/10
Distance covered in the tests [m]	520	920	1450
Delay in velocity profiles [m/s²]	0.8	0.8	0.8

Table 4. Characteristics of the SORT driving tests [9]

4. Results and Analysis

4.1. SORT driving cycles

SORT 1, 2 and 3 driving tests were performed for city buses fueled with diesel and CNG fuels. Since the tests were of comparative nature, the driving conditions of the tested buses were analyzed. In Figures 9–11 a comparison of the speeds of the vehicles in individual tests has been presented along with the coefficients of determinacy that are a measure of reproducibility of the driving parameters. The values of the obtained coefficients fall in the range between R^2=0.994 and 0.999. It is commonly accepted that coefficient R^2>0.95 indicates a significant correlation between the tested parameters. This condition was fulfilled for all driving tests in the presented investigations, the driving parameters were comparable, which renders the comparison of the results of the trials fully justified.

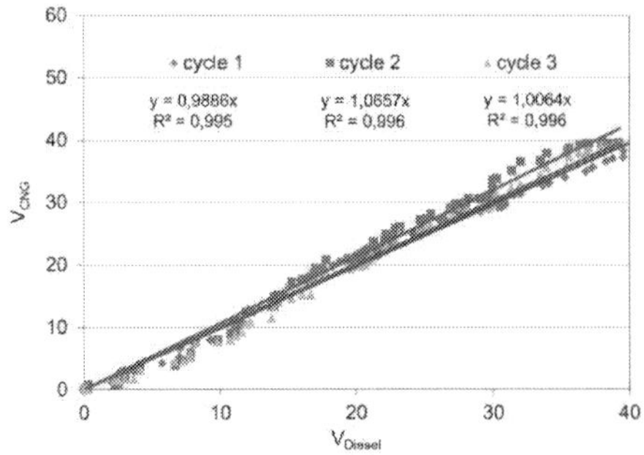

Figure 9. Comparison of speeds of vehicles in subsequent trials of SORT 1 cycles(time compliance allowed for).

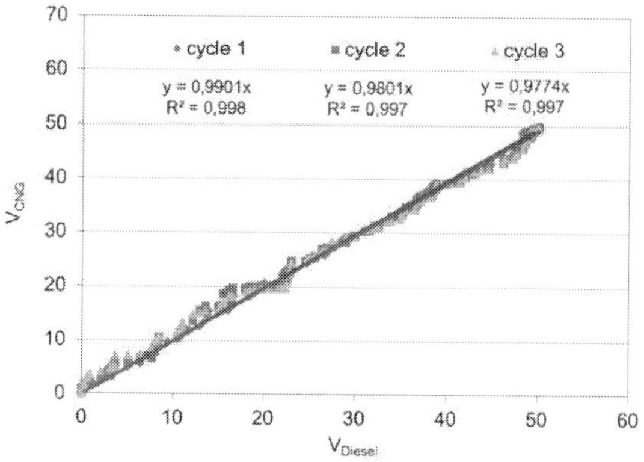

Figure 10. Comparison of speeds of vehicles in subsequent trials of SORT 2 cycles (time compliance allowed for).

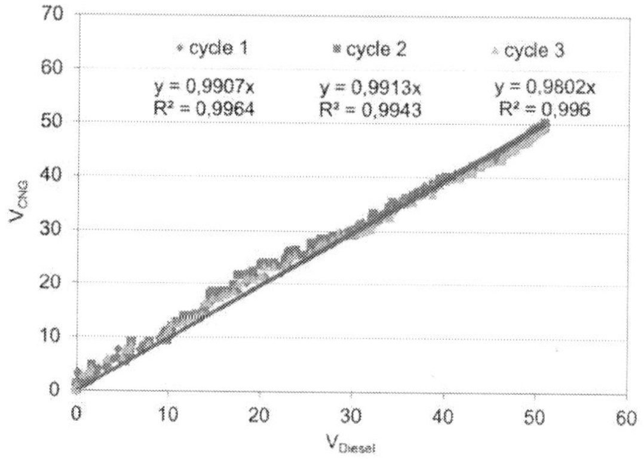

Figure 11. Comparison of speeds of vehicles in subsequent trials of SORT 3 cycles (time compliance allowed for).

Upon analysis of the emission rates it was observed that their greatest values for all emission components occurred in the time when the buses accelerated. It is a result of dynamic changes of engine operating parameters in the acceleration phase. During acceleration, very dynamic changes of engine speed and load take place, reaching momentary values close to maximum, which translates into momentary increase in the exhaust emissions. The greatest values of second-by-second emission of CO for the tested vehicles occurred for drive off and acceleration in the SORT 1 test (Figures 12–14). Momentary acceleration values had a significant influence on the CO emission rate, which is visible in each SORT test. In all analyzed driving tests, the CNG-fueled bus had a higher momentary CO emission rate (up to 307 mg/s) in the acceleration phase. The CO emission rate reduces significantly when the bus speed stabilizes.

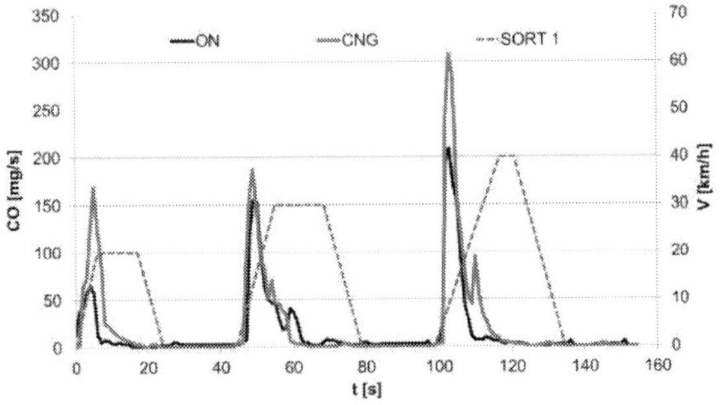

Figure 12. CO emission rate in the SORT 1 test along with the assumed velocity profile.

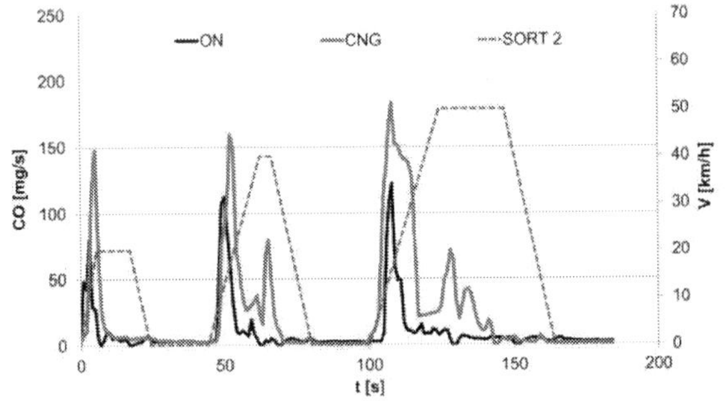

Figure 13. CO emission rate in the SORT 2 test along with the assumed velocity profile.

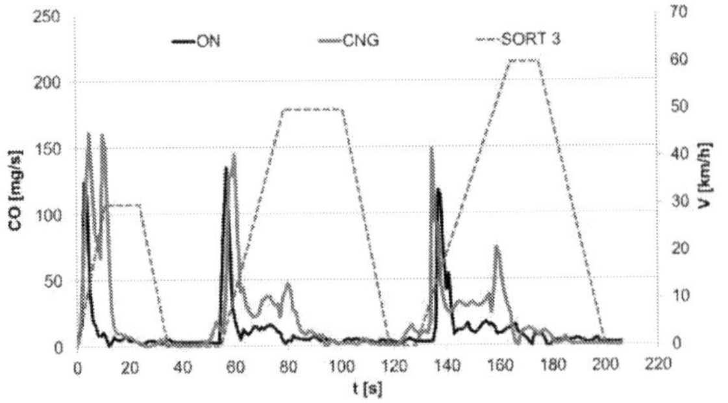

Figure 14. CO emission rate in the SORT 3 test along with the assumed velocity profile.

The emission rate of HC changes analogically to that of CO, higher values were recorded for the CNG-fueled bus (Figures 15–17). The acceleration phase also contributed to the HC emission rate – at this time the emission rate was several times higher than for a constant speed. It is also characteristic that the emission rate of the CNG-fueled vehicle is definitely higher compared to the conventionally fueled vehicle. In the SORT 1 test, the maximum values of the HC emission rate for the CNG-fueled vehicle were 49.2 mg/s, and for diesel- fueled vehicles 1.3 mg/s. For both emission rates (CO and HC) we can see changes resulting from the cooperation of the engine with the transmission. Momentary surges of the emission rates upon reaching the maximum values result from the cooperation of the engine with the transmission system.

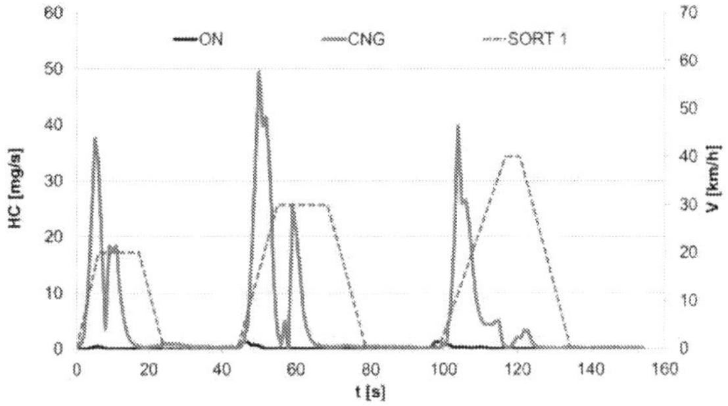

Figure 15. THC emission rate in the SORT 1 test along with the assumed velocity profile.

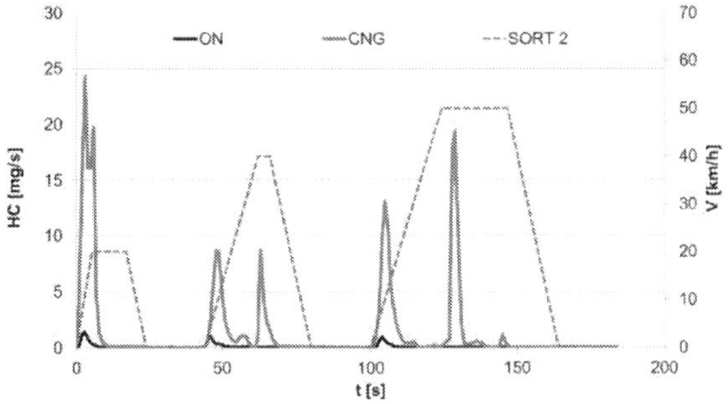

Figure 16. THC emission rate in the SORT 2 test along with the assumed velocity profile.

Very large differences between the CNG- and diesel-fueled vehicles occurred for the emission of NO_x (Figures 18–20). The greatest values were obtained for the diesel -fueled bus in the SORT 1 test. This value reached 302 mg/s during the last stage of acceleration. For the CNG-fueled bus, the NO_x emission rates are much lower (at some points of the SORT tests even tens of times lower) than the values obtained for the diesel- fueled bus. We should thus infer that the value of this emission was significantly influenced by the aftertreatment systems. In the

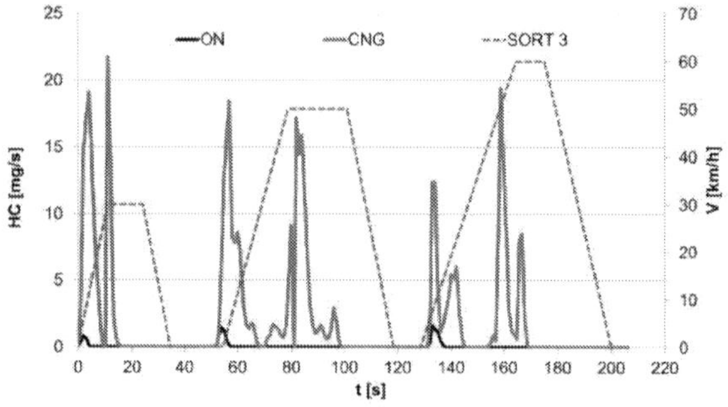

Figure 17. THC emission rate in the SORT 3 test along with the assumed velocity profile.

conventional system, Selective Catalytic Reduction was applied, whose conversion rate heavily depends on the exhaust mass flow (the conversion rate is lower at greater mass flow). The CNG-fueled bus operated with a three -way catalyst, whose conversion rate is not as heavily dependent on the exhaust mass flow as it is on the excess air coefficient.

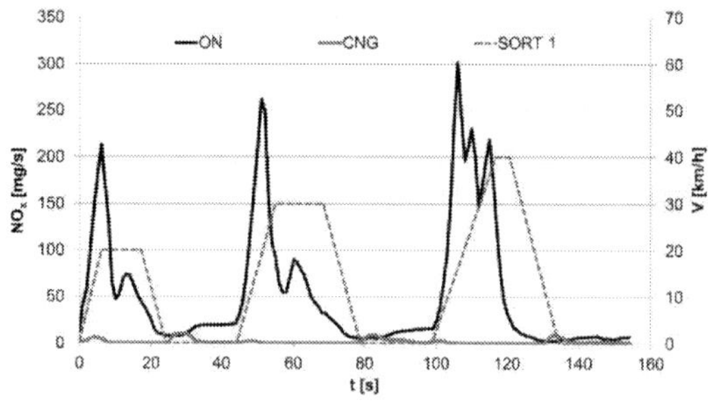

Figure 18. NO_x emission rate in the SORT 1 test along with the assumed velocity profile.

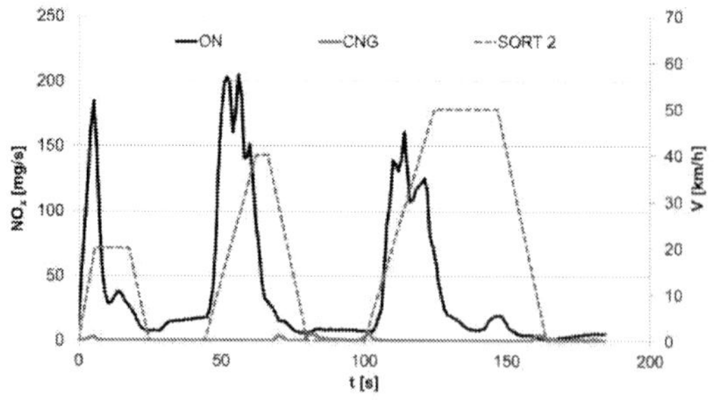

Figure 19. NO_x emission rate in the SORT 2 test along with the assumed velocity profile.

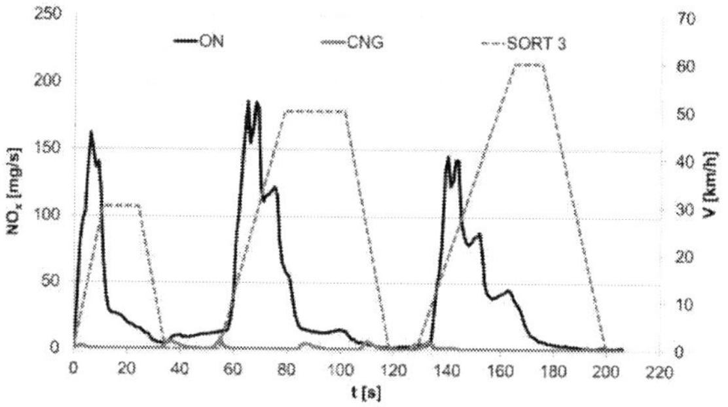

Figure 20. NO_x emission rate in the SORT 3 test along with the assumed velocity profile.

The on-road emissions show similar trends in all SORT tests. From the comparison of these emissions we know that higher values of CO, HC and CO_2 were obtained for the CNG-fueled bus (Figures 21–23). As for CO and HC, it is mainly the effect of engine operation at stoichiometric mixture, which contributes to a greater share of incomplete combustion. It is noteworthy that three-way catalysts have little efficiency in the oxidation of methane one of the main CNG components. Oxidation of methane in catalytic converters requires high temperatures that are usually unobtainable under regular urban traffic operation (hence the lower conversion rate of methane leading to a higher emission of HC for the CNG-fueled bus). It is the HC emission differences that were the highest in all the driving tests (in SORT 2 and 3 they were over 20 times higher for the CNG-fueled bus compared to the diesel- fueled one). The differences in the road emissions of CO and NO_x were lower than those of HC. From the comparison of individual components, we know that the CNG-fueled bus had higher emissions of CO_2 (fuel consumption) in all driving tests. These were small differences. The greatest differences in the SORT 1 test reached 15%. Given the much lower price of CNG compared to diesel fuel, this fuel is far more interesting for the users.

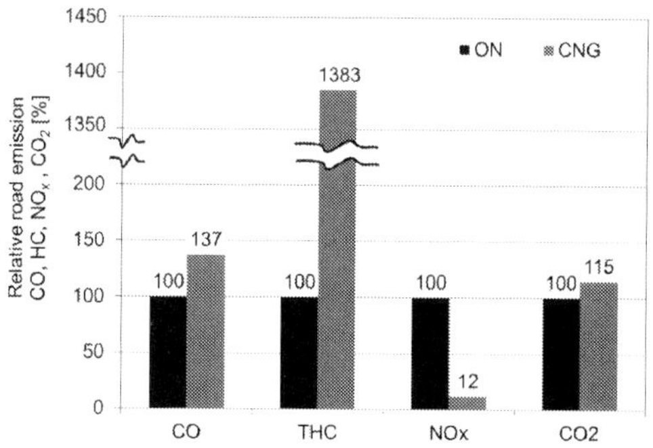

Figure 21. Comparison of relative road emissions in the SORT 1 test.

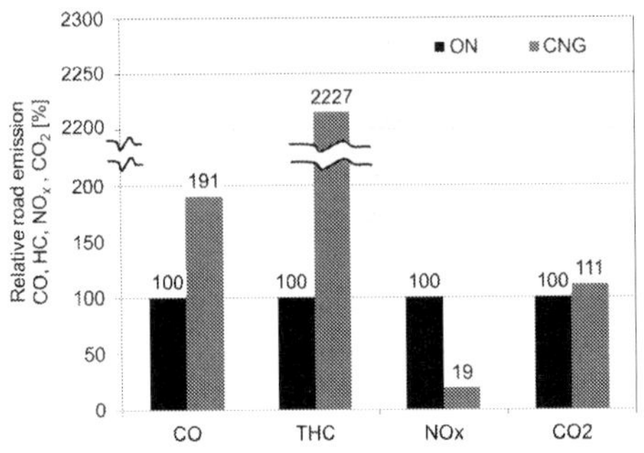

Figure 22. Comparison of relative road emissions in the test SORT 2 test.

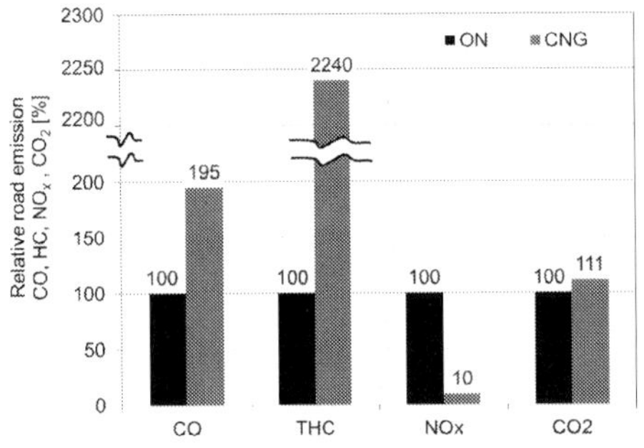

Figure 23. Comparison of relative road emissions in the SORT 3 test.

5. Tests under actual traffic conditions

Another stage of the investigations were measurements performed under actual traffic conditions. In order to reflect true operating conditions, regular passengers were carried during the tests and the buses serviced an actual route (Figure 7). The number of vehicle stops resulting from the process of passenger carriage was preserved. It is, however, difficult to reproduce identical parameters of the vehicle drive due to a variety of factors independent from the vehicle or the driver. One of such factors is the traffic congestion. For the compared drives an analysis of speed changes was performed (Figure 24). From this comparison we know that the speeds of the CNG and the diesel- fueled buses differed. The average speed of the diesel- fueled bus was 18.9 km/h (maximum 68.9 km/h), and the CNG-fueled bus 19.8 km/h (maximum 57.3 km/h). Average speeds, which are the most objective parameter of the assessment of the two drives differed slightly (by 1 km/h). We can thus assume that the comparison of the obtained results of the exhaust emission tests was justified.

Using the data pulled from the CAN network, time density characteristics of the tested engines were developed (Figure 25). The diesel- fueled engine most frequently operated in the speed range up to 1600 rpm. There existed two most frequent load ranges 0–400 Nm and 1000–1400 Nm. The share of the work area when the engine operated at high idle (up to 1600 rpm) is also significant and constituted 18% of the total test time. For the CNG-fueled engine two areas were most frequently used. The first one (of the largest operating time share) occurred for the engine speeds up to 800 rpm and loads up to 200 Nm and, which was 35% of the operating time. The other fell in the range from 800–1200 rpm in the entire load range constituting 49% of the total operating time. It can thus be confirmed that it predominantly operated under load characteristics.

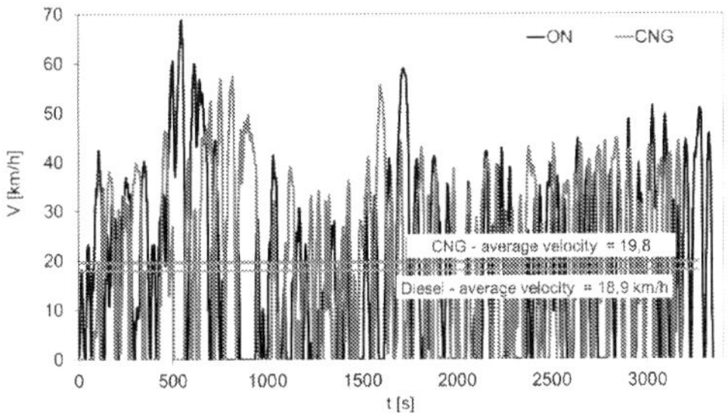

Figure 24. Velocity profiles of CNG and diesel- fueled buses obtained under actual traffic conditions on the city route.

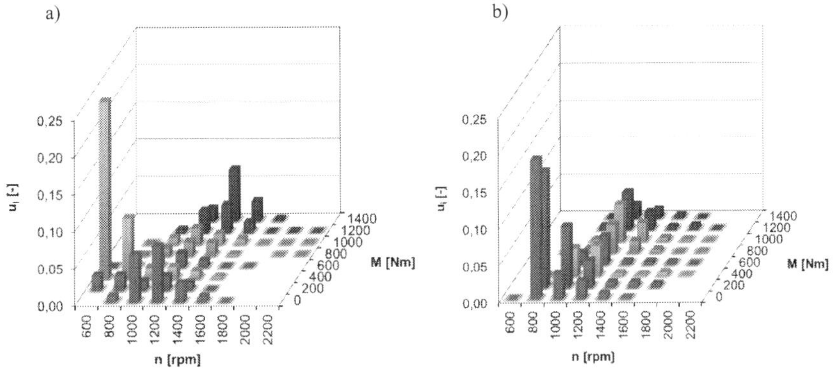

Figure 25. Time share characteristics of the engine speed and engine load during the test on the city route: a) diesel- fueled vehicle, b) CNG- fueled vehicle.

Based on the research performed under actual traffic conditions the values of road emissions were determined and compared (Figure 26). Similarly to the SORT tests, the CNG- fueled vehicle was characterized by a higher emissions of CO, THC and CO_2. For the mentioned exhaust components the differences between the CNG and diesel- fueled vehicles were 78%, 1843% and 18% respectively (Figure 27). The emission of NO_x from the CNG bus was lower by 87%. The specificity of the city route allows for the urban traffic conditions and the obtained results indicate that none of the performed SORT tests fully reflects the parameters characteristic of this route. The obtained exhaust emissions values were influenced by the properties of the applied fuels and the type of thermodynamic cycle of the tested engines.

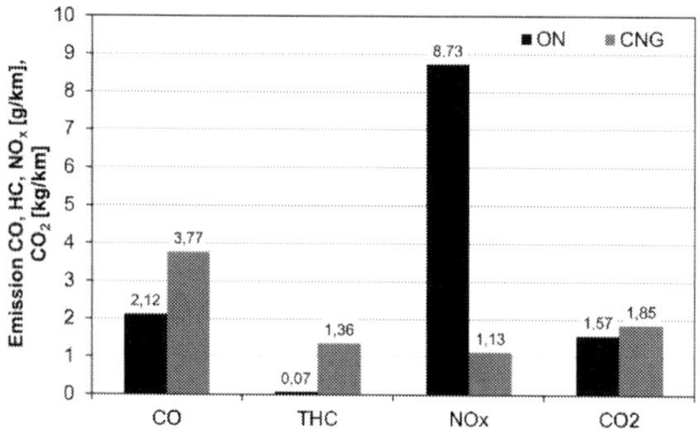

Figure 26. Comparison of the relative road emission obtained in the tests on the city route.

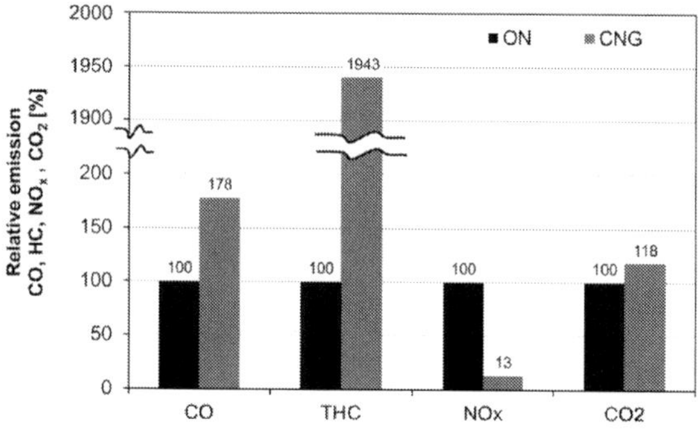

Figure 27. Comparison of the relative road emission obtained in the tests on the city route.

6. Summary

The conducted tests and their results confirm that CNG fuel is a viable alternative to diesel fuel. The results of emission of toxic compounds indicate a high potential for the use of CNG to power city buses and other vehicles. However, the emissions of HC and CO from CNG engines remain a problem. In case of HC emission, the key element is methane, which is a compound difficult to burn even in the oxidation catalysts. Thus, the development of catalytic converters is still an important issue; this should be seen primarily as a possibility of reducing the HC and CO emissions, of course in combination with the optimization of the combustion process. A positive aspect of the presented test is less NO_x emissions from the CNG engine when compared to a diesel engine. Taking into account the aftertreatment emission control methods of NO_x reduction is still the most problematic process. CNG buses, offering low PM and NO_x emissions, should be regarded as eco-friendly means of public transport. In addition to environmental concerns, the economic considerations should also be taken into account and currently CNG is a much cheaper fuel than diesel. Another argument for using CNG to power vehicles is the availability of this fuel. According to estimates Poland alone holds shale gas resources that may even equal 5.4 trillion m^3.

Author details

Jerzy Merkisz*, Michał Dobrzyński, Miłosław Kozak, Piotr Lijewski and Paweł Fuć

*Address all correspondence to: Jerzy.Merkisz@put.poznan.pl

Poznan University of Technology, Institute of Combustion Engine and Transport, Poznan, Poland

References

[1] Autogas in Europe, The Sustainable Alternative – An LPG Industry Roadmap. Report of European LPG Association, Brussels 2009.

[2] Budzik G.: Engine power supply for city buses with compressed natural gas. Publishing House of Rzeszów University of Technology, Rzeszów 2006.

[3] Dobrzyński M.: The impact of the use of natural gas on the environmental performance of selected modes of transit. PhD thesis. Poznań University of Technology, Poznań 2015.

[4] Serdecki W. [editor]: The study of internal combustion engines. Publishing House of Poznań University of Technology, Poznań 2012.

[5] Kowalewicz A.: Combustion systems for high-speed internal combustion engines; Transport and Communication Publishers, Warszawa 1990.

[6] Merkisz J.: Ecological problems of internal combustion engines. Publishing House of Poznań University of Technology, Poznań 1999.

[7] Jeuland N., Montagne X.: European Emission Test Programme (EETP) – Final Report. Institut Français Du Petrole 2004.

[8] Merkisz J., Kozak M.: Meeting the new exhaust emissions standards for engines powered by gaseous and conventional fuels. V International Scientific Conference Gas Engines 2000, Vyšné Ružbachy, Slovakia, 8-12.05.2000.

[9] UITP – International Association of Public Transport: SORT – Standardised On-Road Test Cycles, Brussels 2003.

Chapter 2

Rapeseed Oil Methyl Esters (RME) as Fuel for Urban Transport

Jerzy Merkisz, Paweł Fuć, Piotr Lijewski and Miłosław Kozak

Additional information is available at the end of the chapter

http://dx.doi.org/10.5772/62218

Abstract

The use of biofuels is justified by the common agricultural policy decisions, by the need to improve environment protection and by the search of alternative energy sources. In such a context, methyl esters of vegetable oils, known as biodiesel and ethyl alcohol are receiving increasing attention as alternative fuels for automotive engines. The main advantages of biodiesel and ethyl alcohol are that these fuels are nontoxic, biodegradable, and renewable with the potential to reduce engine exhaust emissions, especially with regard to greenhouse gases emission. The fact that these biofuels are available in larger and larger quantities is of great importance as well. Currently, in the European market the most important biofuel is FAME (Fatty Acid Methyl Esters) manufactured mainly as Rapeseed Methyl Esters (RME). It is forecasted that the scale of production and consumption of this fuel will continue increasing as a result of the growing demand for diesel fuels and a levelled demand for spark-ignition engine fuels. Currently, FAME is added to regular diesel fuels in the amount of up to 7%. Besides, its consumption in a pure form grows as well. This chapter presents ecological properties of RME in relation to conventional diesel fuel. The aim of the research was to determine the potential of RME in reducing exhaust emissions (CO, HC, NO_x and PM) from diesel engines operated in buses. The tests were carried out in real operating conditions of a city bus meeting EEV emissions standard. Comparative analysis made it possible to assess the environmental performance of buses depending on the type of fuel used. The obtained results indicate a slightly lower emission of CO, HC and PM when the vehicle was fuelled with RME but at the same time its application results in a slight increase in the emission of NO_x. It seems that similar level of exhaust emissions recorded regardless of fuel type results from an advanced exhaust gas aftertreatment system (SCR + DPF) which was applied in the test vehicle.

Keywords: RME, exhaust emissions, real drive emissions (RDE), city bus

1. Introduction

Currently, the most important factor stimulating the advancement of vehicle powertrains and fuels are environmental aspects. This is a result of increasingly stringent exhaust emissions legislation. The dominating source of fuels is still crude oil. However, a slow but steady growth in the share of biofuels in the market has been observed. This is mainly owing to the appropriate policy that imposes on the manufacturers and distributors the obligation to increase the share of biofuels in the world fuel market each year.

As replacement fuels for combustion engines, several synthetic, mineral or plant-based substances as well as their combinations are taken into consideration. The real alternatives for crude oil-based fuels are only those that:

- Are available in sufficient amounts,
- Are characterized by technical and energy-related properties that determine their applicability in combustion engines,
- Are cheap in production and distribution,
- Are less hazardous to the environment than conventional fuels,
- Ensure acceptable economic indexes of engines and safety of use.

The application of renewable fuels (biofuels in particular) specifically aims at protecting Earth's natural resources, reducing the emission of carbon dioxide and liberating from fossil fuels. The application of renewable fuels also positively influences vehicle exhaust emissions, i.e. air pollution.

Currently, the most important alternative fuel in the European market is FAME (*fatty acid methyl esters*), manufactured as methyl esters of fatty acids – RME (*rapeseed methyl esters*). It is forecasted that the scale of production and consumption of this type of fuel will continue to grow due to a growing demand for diesel fuels and a stable demand for gasoline. Today, FAME is added to conventional diesel fuels in the amount of 7%. Its use in its pure form also grows.

Plant-based oils suitable for production of fuels are obtained not only from rapeseed but also from sunflower, soy, peanuts, oil palm, linseed or hemp. The history of application of plant-based oils for fuelling of combustion engines dates back to the times of the first diesel engines – Rudolf Diesel used peanut oil for his engines. Subsequently, owing to a growing accessibility to cheap diesel fuel, the idea of application of plant-based oils was abandoned until the fuel crisis in the 1970s of the past century.

Direct replacement of diesel fuel with plant-based oils in standard diesel engines does not give entirely positive results because of the rapid formation of deposits on the surfaces of pistons, piston rings and injectors. Another downside is high viscosity that limits the use of plant-based oils to the temperature above approx. 10°C (otherwise fuel preheat is necessary).

Fatty acids methyl esters, as a fuel, are seen in a much better light, whose physicochemical properties are close to diesel oil (Table 1). Rapeseed oil methyl esters may be manufactured industrially or in small processing facilities (agro-refineries). Industrial production methods

http://dx.doi.org/10.5772/62218

utilize hot technologies requiring the reaction of transesterification at the temperature of approx. 240°C under the pressure of approx. 10 MPa as well as high surplus of methanol returned to the process. In non-industrial technologies, the esterification process is realized at a much lower temperature: 20–70°C, under atmospheric pressure and with a lower surplus of alcohol but in the presence of an alkaline catalyst. Transesterification continues according to the following plan:

$$100\ kg\ of\ rapeseed\ oil + 11\ kg\ methanol \rightarrow$$
$$\rightarrow 100\ kg\ methyl\ esters + 11\ kg\ glycerin$$

The technological and operational benefits arising from the application of FAME as fuel are as follows [1]:

- High cetane number → possibility of achieving high engine speeds and injection delays (reduction of NO_x formation rate),
- The fuel has good lubrication properties, may be used as a lubrication additive in low sulfur diesel fuels,
- Little toxicity and irritation to human body (no polycyclic aromatic hydrocarbons),
- Good biodegradability (pure FAME),
- Reduction of the emission of CO, HC, PM, SO_2 as well as the smoke level,
- In particulate matter, a lower number of insoluble fraction (INSOL) is found,
- The emission of CO_2 is reduced (partial closure of the CO_2 cycle),
- Low sulfur content → lesser exposure of the aftertreatment systems to sulfur,
- Good cooperation with oxidation catalysts and DPFs – the efficiency of these devices is higher compared to pure diesel fuel; no long-term research results available however,
- Reduced engine noise,
- Relatively high ignition temperature → safety in operation.

The disadvantages and risks resulting from the application of FAME as engine fuel are as follows:

- Lower calorific value → higher fuel consumption (by approx. 8–14% for pure esters),
- Reduced vehicle acceleration (up to 10%),
- Possible increased emission of NO_x (up to 15%),
- Increased emission of aldehydes,
- Higher viscosity → impact on the fuel atomization and maximum fuel pressures,
- Higher elasticity coefficient → increased fuel injection pressure,

- Worse low-temperature properties, significant increase in viscosity → difficult engine start at low ambient temperatures, possible fuel pump failure,
- Increased lubricating oil dilution; the penetrating esters lead to precipitation of deposits on oil sump and crankcase, shorter oil change interval,
- Engine oil cooperating with ester fuel is characterized by a reduced capability of dispersing deposits,
- Reduces durability of components made from elastomers and rubbers when in contact with the fuel; sensitive materials: nitrile rubber, polypropylene, nylon and resistant materials: PTFE and viton,
- Corrosion of paint layers in contact with the fuel,
- Strong deposit formation-related corrosive effect on alloys containing copper, certain corrosive effect on steel, aluminium, zinc and lead,
- Intense hygroscopy – the fuel is capable of bonding 40 times more water than diesel fuel,
- Low resistance to hydrolysis; under the influence of water, the esters hydrolyze to acids (corrosive effect) and alcohols, sludge and precipitations occur that may block the fuel filter,
- Greater susceptibility to microbiological contamination, biocide application recommended,
- Residual presence of catalyst in the fuel → blocking of injection nozzles,
- Possible presence of methanol in the fuel → intensification of corrosion, reduced ignition temperature of fuel,
- Possible presence of glycerine in the fuel → corrosion of non-ferrous metals, cellulose fuel filters absorbing glycerine, deposits on moving components of the fuel pump,
- Worse thermo-oxidation stability, rapid deterioration of fuel properties when stored, storage longer than 5 months not recommended,
- Little data on long-term (thousands of kilometres) influence of esters on the engine durability and operation,
- Unexplored results of the influence of theses fuels on modern engines.

2. Application of FAME in urban transport

The use of ecological fuels is particularly appropriate in urban areas where large numbers of people are threatened by automotive pollution. According to estimates by the International Association of Public Transport, 50% to 60% of public transport in Europe is done using buses. Only about 5% of these buses is powered by fuel other than conventional (diesel) [13].

FAME fuel is one of the most often used alternative fuels for vehicles, and it is used in its pure form or as mixtures with diesel fuel. Among the 131 examples of the use of alternative fuels and propulsion units in transport systems of European urban centers analyzed in the framework of the European ALTER-MOTIVE project (Deriving effective least-cost policy strategies for alternative automotive concepts and alternative fuels), 12 cases are related to the use of FAME, including three cases of powering city buses by neat FAME [14]. About 400 buses powered by pure FAME are utilized in Stockholm. There are 27 buses powered by FAME utilized in Burgos (Bulgaria) and 20 in San Sebastian (Spain). In Paris, a decision was made to use 30% RME additive to diesel fuel. About 300 buses are powered in this way (about 7% of the total fleet). The same FAME additive is also used in Valencia. Graz (Austria) also has extensive experience in the use of buses powered by FAME, whereas Rotterdam and Dublin withdrew from the use of this fuel [15].

Parameter	**Gasoline**	**Diesel fuel**	**Ethanol**	**Rapeseed oil**	**Rapeseed biodiesel**
Octane number (RON)	95	20	109	–	8
Cetane number	12	50	8	34	56
Ignition temperature [°C]	< 0	60	13	285	168
Density at 20°C [kg/m^3]	730	820–860	789	920	880
Kinematic viscosity at 20°C [mm^2/s]	–	2.8–5.9	1.4	76	6.9–8.2
Cold filter blocking temperature [°C]	–	winter grade: –20	–	20	-7
Calorific value					
[MJ/kg]	43.5	43	26.8	37.4	37–39
[MJ/dm^3]	31.8	36	21.2	34.4	33.5–34.3
Range of distillation temperatures [°C]	40–215	180–380	78.3	–	320–360
Sulfur content [ppm]	<10	<10	–	–	<10
Air demand [kg/kg]	14.3	14.5	9.0	–	12.7
Elemental composition [%]					
C	84.8	87	52.2	–	77
H	13.3	13	13.0	–	12.5
O	2.4	0	34.8	–	10.5

Table 1. Comparison of the properties of conventional engine fuels and selected biofuels [1]

As already mentioned, the application of FAME as a fuel in its pure form or as an additive may reduce the exhaust emissions from a diesel engine. The reduction of the emissions of CO, HC and PM and an increase in the emissions of NO_x is most frequently observed in the case of this fuel [2, 3]. Literature presents many works related to this subject [4–12]. It is noteworthy, however, that much of the research works treating on FAME were based on investigations of engines of older generations (utilizing conventional injection systems). This is partly due to a recent dynamic advancement of diesel engines (new injection systems – huge increase in the injection pressure, downsizing, etc.). Besides, many of the said works are based on measurements for only one or several points of engine work, which provides only a limited view on the influence of FAME on the exhaust emissions in the entire area of operation. It is also noteworthy that most of the investigations described in literature were performed on passenger vehicles. The above-mentioned issues were the main reason for the initiation of the research described in this chapter. The investigations covered the measurement of the exhaust emissions from a modern city bus under actual traffic conditions, fuelled by diesel fuel and RME for comparison.

3. Methodology

The investigations were carried out under actual vehicle operating conditions (RDE). The object of the tests was an 18 m city bus. The tests began with a run on diesel fuel and then the runs were repeated for RME (rapeseed methyl ester, commercial name: B100). The bus was fitted with a combustion engine of the displacement of 8.9 dm^3 and the power output of 231 kW. The engine aftertreatments were SCR and DPF. Basic technical specifications of the tested bus have been given in Table 2, and Figure 1 shows the test object ready for the test runs.

Ignition	Compression ignition
Displacement	9.2 dm^3
Number of cylinders	6
Arrangement of cylinders	Straight
Compression ratio	17.4
Maximum power	231 kW at 1900 rpm
Maximum torque	1275 Nm at 1100–1700 rpm
Emission technology	EEV
Aftertreatment	SCR + DPF
Length	18 000 mm
Height	3 400 mm
Vehicle weight	24 000 kg

Table 2. Basic parameters of the tested vehicle and its engine

http://dx.doi.org/10.5772/62218

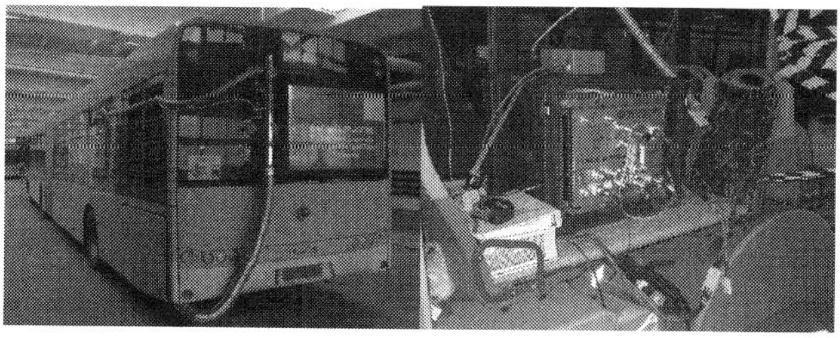

Figure 1. City bus with the measurement equipment fitted during the road tests

In the on-road city bus exhaust emissions tests, a PEMS portable exhaust emissions analyzer was used:

- Semtech DS by Sensors Inc., measuring and recording:
  - The concentrations of CO and CO_2 (NDIR analyzer – non-dispersive infrared), $NO_x = NO + NO_2$ (NDUV analyzer – non-dispersive ultraviolet), HC (FID analyzer – flame ionization detector), O_2 (electrochemical sensor);
  - Thermodynamic exhaust gas parameters (mass flow, temperature, pressure) – the mass flow utilizes the Pitot tube;
  - Ambient conditions– ambient pressure, temperature, humidity;
  - Vehicle position and speed – GPS system;
  - Data from the vehicle on-board diagnostic systems – data transmission protocol CAN SAE J1939/J2284.

The analyzed exhaust gas sample was taken from the mass flow meter and carried via a heated line maintaining the temperature of ~190°C (Figure 2). This aimed at preventing HC condensation on the duct walls. Upon passing the filter, the sample reached the FID analyzer, where the concentration of HC was measured. Upon chilling to the temperature of 4°C, the sample was directed to the NDUV and NDIR analyzers. These analyzers measured $NO_x = (NO + NO_2)$, CO and CO_2. At the end, the electrochemical sensor measured the concentration of O_2. A portable computer paired to the main unit via WIFI realized the control and monitoring of the Semtech DS. The systems can communicate via LAN network, yet in these investigations this way of communication was not utilized.

- AVL Micro Soot Sensor for the measurement of PM. This analyzer utilizes a photo-acoustic measurement method that consists in radiating of the particles with modulated light, which leads to their intermittent heating and cooling. In this way, the carrier gas intermittently changes its volume, acting like a sound wave. The measurements utilize microphones that are sensitive to vibrations only in a specified frequency and amplitude range. When the air is clean, no signal is detected; but when the particle number increases in the gas (increased concentration), the value of the sound signal increases. In order to avoid condensation, the soot in the exhaust gas is diluted.

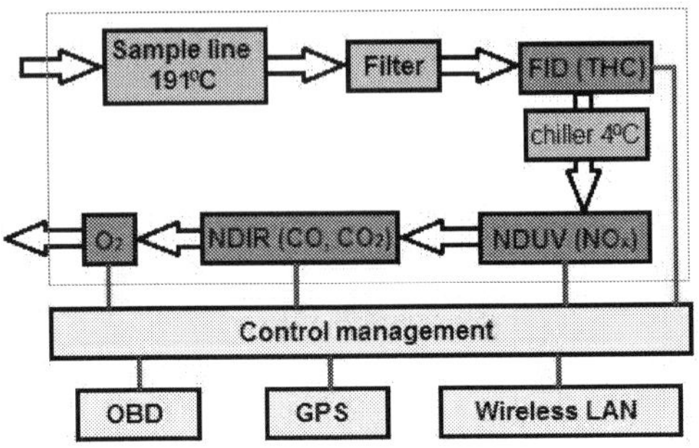

Figure 2. Diagram of a portable exhaust emissions analyzer SEMTECH DS; exhaust gas flow channels (arrows) and electrical connections circled (blue line)

The measurements of the exhaust emissions were performed in the SORT tests and in actual traffic, when the buses operated on a regular line in the city of Poznań. The selected line reflected traffic conditions typical of city bus operation in urban agglomerations (Figure 3). The selected city line is classified by the operator (MPK) as one of the most heavily loaded in terms of passenger count and is one of the longest (its length is 13.1 km). The line includes 30 bus stops. A varied configuration of the test line (main roads, residential area passages and downtown areas) provides a high variability of accelerations and a high share of road congestions, which enables analysis of exhaust emissions in a wide range of vehicle-operating parameters.

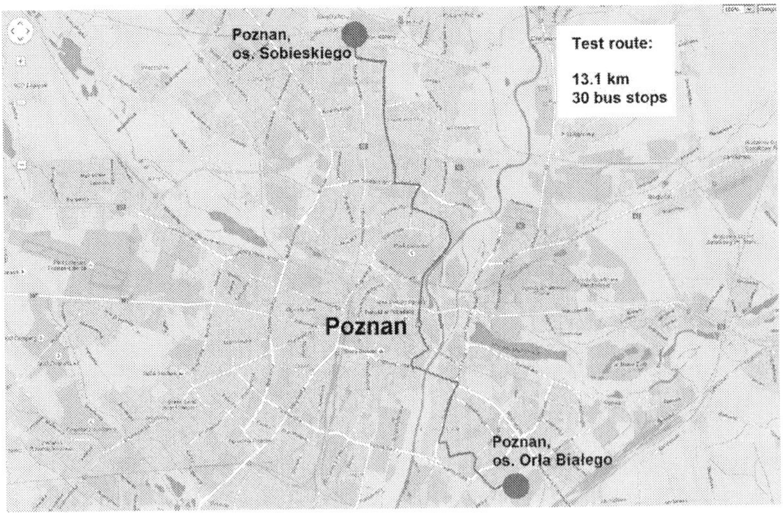

Figure 3. The RDE test route (created by gpsvisualizer.com)

The second stage of the tests covered the SORT runs that are the universal and commonly accepted method of assessment of fuel mileage (mainly) and exhaust emissions. These tests are divided into segments and are a representation of three types of traffic – downtown routes, general urban routes and extra urban routes (Figure 4, Table 3). The basic module of the SORT cycle is described with driving speed, length of route and driving time. These parameters create a velocity profile characteristic of a given route that includes stops at intersections, bus stops, driveoffs and cruise.

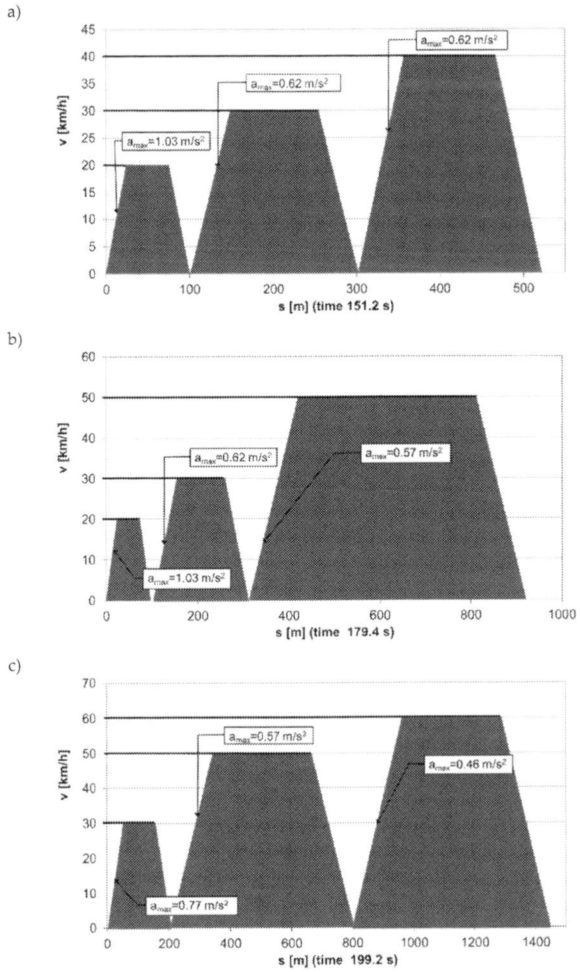

Figure 4. SORT test velocity profiles: (a) SORT 1, (b) SORT 2, (b) SORT 3

	SORT 1	SORT 2	SORT 3
Average speed [km/h]	12.6	18.6	26.3
Share of stopped vehicle in the test [%]	39.7	33.4	20.1
Constant speed in profile 1 [km/h]/[m]	20/100	20/100	30/200
Acceleration in profile 1 [m/s²]	1.03	1.03	0.77
Constant speed in profile 2 [km/h]/[m]	20/200	40/220	50/600
Acceleration in profile 2 [m/s²]	0.77	0.62	0.57
Constant speed in profile 3 [km/h]/[m]	40/220	50/600	60/650
Acceleration in profile 3 [m/s²]	0.62	0.57	0.46
Time of stoppage after each profile [s]	20/20/20	20/20/20	20/10/10
Distance covered in the test [m]	520	920	1450
Delay in velocity profiles [m/s²]	0.8	0.8	0.8

Table 3. Characteristics of the SORT tests

4. Results and analysis

4.1. The SORT tests

Measurements were performed in the SORT 1, 2 and 3 tests for the bus fuelled with diesel oil and B100. The performed analysis of results was of a comparative nature and its aim was to present the exhaust emissions for diesel and B100 fuels. Such a policy was adopted not only for the exhaust emissions tests under the SORT test conditions but also for the measurements performed under actual traffic conditions, on an actual bus line – this analysis is described in the further part of this chapter. In a comparative analysis, a very important factor is the reproducibility of conditions and parameters in comparable tests, which is why the first part of the results analysis is related to the reproducibility of the bus driving parameters in the SORT test. Figures 5, 6 and 7 present the velocity profiles of a bus fuelled with diesel fuel and B100 in the SORT 1, 2 and 3 tests, respectively. The obtained velocity profiles deviate from the reference ones (Figure 4) but it is admissible, though attention must be paid to the reproducibility of the test runs of the bus fuels with diesel fuel and B100. The presented velocity profiles confirm the reproducibility of the tests and the observed miniscule differences (less than 5%) did not significantly influence the investigations. It can, thus, be assumed that the conditions of measurement and driving parameters of the bus in the compared tests were reproducible, which fully justifies the comparative analysis.

The SORT tests showed a trend of a slight reduction of the emission of PM, HC and CO when the bus was fuelled with B100. The exhaust emissions test results have been presented in Figures 8, 9 and 10. In all three tests, a reduction in the emission of PM was achieved, the greatest – 9% in the SORT 1 test. The application of B100 resulted in the reduction of the

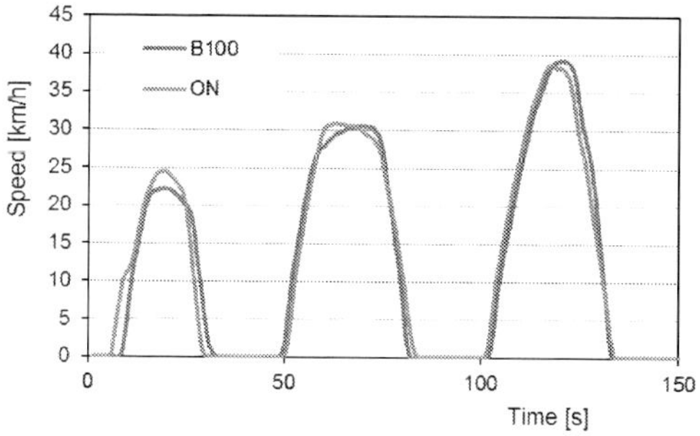

Figure 5. Comparison of the bus speeds in the SORT 1 test

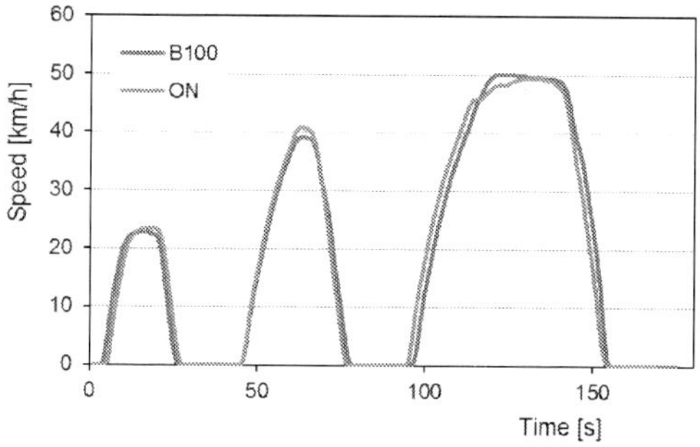

Figure 6. Comparison of the bus speeds in the SORT 2 test

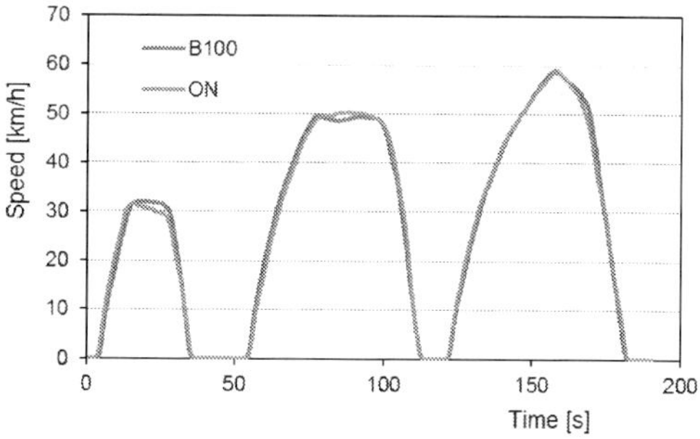

Figure 7. Comparison of the bus speeds in the SORT 3 test

emission of HC in the SORT 2 and 3 tests, but in the SORT 1 test this emission remained unchanged. The results are similar for the emission of CO – the application of B100 resulted in a reduction of this component in the SORT 2 and 3 tests by 16% and 4% respectively but for the SORT 1 test a slight increase in the emissions of CO was recorded (2%). A disadvantageous phenomenon accompanying the use of B100 is increased emission of NO_x observed in the SORT 1 and 3 tests – this emission increased by 13% and 5% respectively. In the SORT 2 test only a slight difference between diesel oil and B100 was observed. This difference was merely 1%. This value is so low that it does not support the trend observed in the SORT 1 and 3 tests. The increase in the emission of NO_x and a simultaneous reduction of the emissions of CO, HC and PM may be caused by the presence of oxygen atoms in the B100 molecules, which influences the course of combustion inside the cylinder. A greater share of oxygen increases the combustion and heat release rates as well as the temperature and availability of oxygen in a combustion chamber, the effect of which is increased emission of NO_x and reduced emissions of products of incomplete combustion.

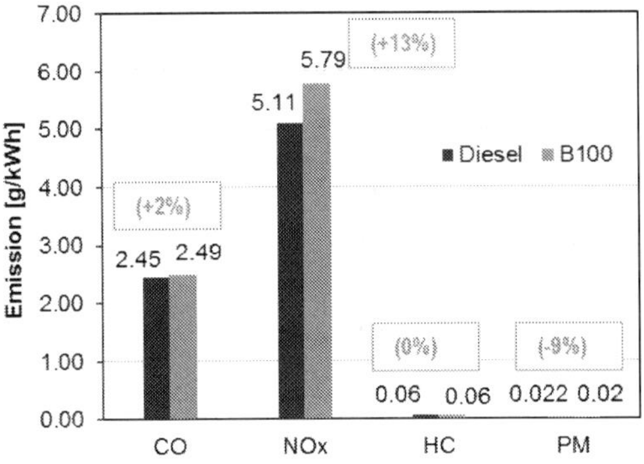

Figure 8. Comparison of relative road exhaust emissions in the SORT 1 test

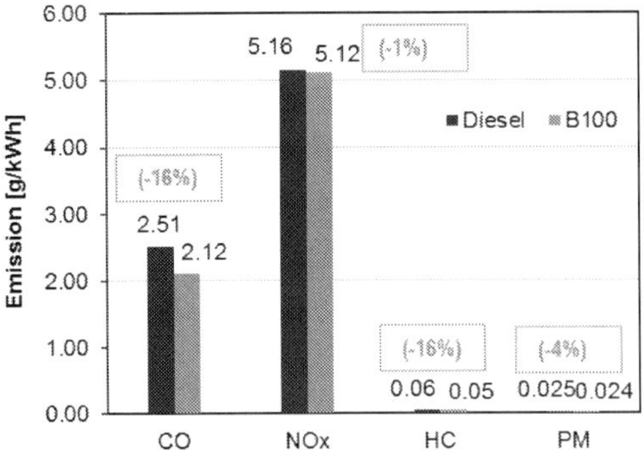

Figure 9. Comparison of relative road exhaust emissions in the SORT 2 test

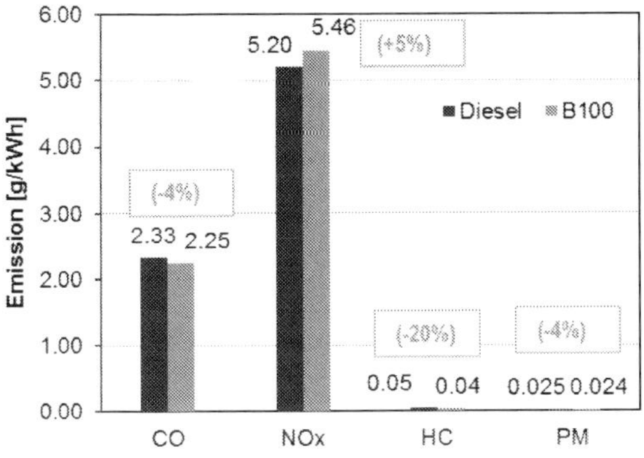

Figure 10. Comparison of relative road exhaust emissions in the SORT 3 test

4.2. Tests under actual operating (traffic) conditions

The second stage of the investigations were measurements performed under actual city traffic conditions. This method of research was selected, as it is relatively new and provides the actual exhaust emissions of the entire vehicle (the method includes all factors occurring during bus operation in real traffic). The HDD engines, including engines fitted in city buses, are mostly tested on engine test brakes. Such tests are incapable of reproducing the actual emissions of actual test runs (it is impossible to ideally reproduce real traffic conditions), which is why the authors decided on tests under actual traffic conditions. It is noteworthy that this type of research becomes increasingly important also in the context of homologation procedures.

Similarly to the SORT tests, the conditions of the test run of the vehicle fuelled with diesel oil and B100 were analyzed in the first place. In order to ensure maximum fidelity of the operating conditions during the tests, passengers were carried during the test run. The number of bus stops was actual, as well as the resultant number of vehicle stops. Despite the fact that the bus covered the same distance on public roads, the runs were characterized by the influence of a variety of unpredictable factors that might have a significant impact on the driving parameters – the exhaust emissions test results. The authors needed to analyze the conditions and parameters of the bus drives for both analyzed cases. The first parameter that was analyzed in this context was the bus speed (Figure 11). In both tests, the nature of the changes of the velocity profile is similar and the difference of the average speeds is small (19.3 km/h for the diesel-fuelled bus and 18.7 km/h for the B100-fueled bus, i.e. 3%). The lower average speed of the B100-fueled bus resulted in the extension of the run time by 230 s, which is a relatively small change of merely 7%. The difference most likely results from the variable traffic conditions (road congestion).

Another element of the comparison of the conditions and parameters of the bus drive for both fuels is the engine operating parameters. Using the data pulled from the CAN, a characteristic presenting the engine torque against engine speed was created (Figure 12). From this characteristic, it results that the engine work areas in both measurement cycles overlap. It is note-

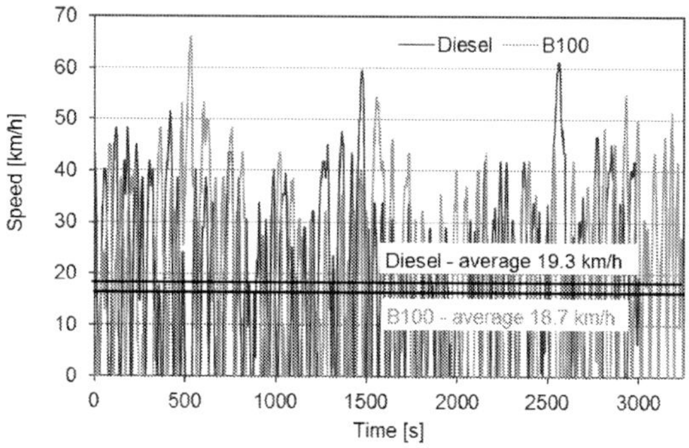

Figure 11. Vehicle speeds during the road tests

worthy that the engine very often worked in full load characteristics and for low loads and engine speeds. This is the case for both the diesel fuel and B100. In order to perform a full analysis also the engine time share characteristics (Figure 13) were created showing the engine operating time share at a point defined by the engine speed and torque. The operating time shares were very similar. The engine most frequently operated in two intervals – the first was 1000–1600 rpm and load 800–1300 Nm. The second most frequent engine operating interval was 600–1400 rpm and load of 0–300 Nm. In both analyzed cases, the operating time share at idle was significant and amounted to 24% when fuelled with diesel fuel and 27% when fuelled with B100. Since the drive and engine parameters were mostly similar, the authors can confirm that the comparison of the obtained results is justified.

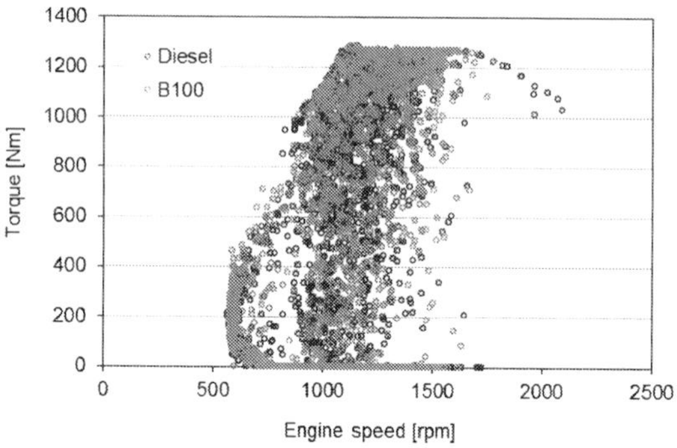

Figure 12. Interval of engine torques and speeds during the road tests

The exhaust emissions determined under actual traffic conditions confirm the trend observed in the SORT tests. The application of B100 fuel resulted in a reduction in the emission of particulate matter by 9% and a reduction in the emission of HC and CO by 14% and 19%

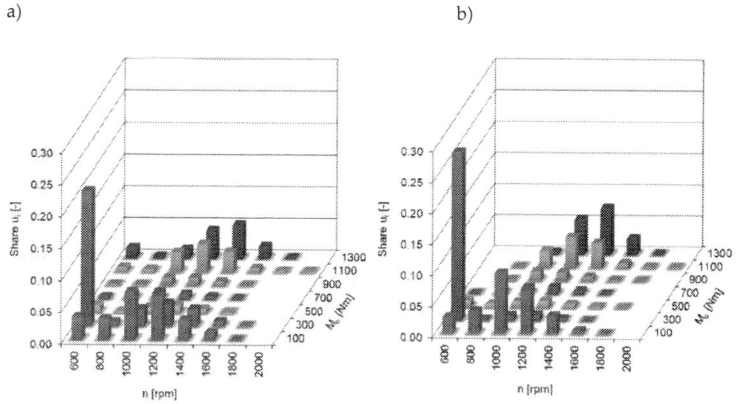

Figure 13. Operating time share characteristics in the engine speeds and load intervals during the road tests on the city line: (a) diesel-fuelled bus, (b) B100-fueled bus

respectively. At the same time, the application of B100 slightly increased the emission of NO_x – by 7% (Figure 14). The cause of such changes is most likely the changes in the process of combustion, resulting from the chemical composition of the fuel, as was the case in the SORT tests.

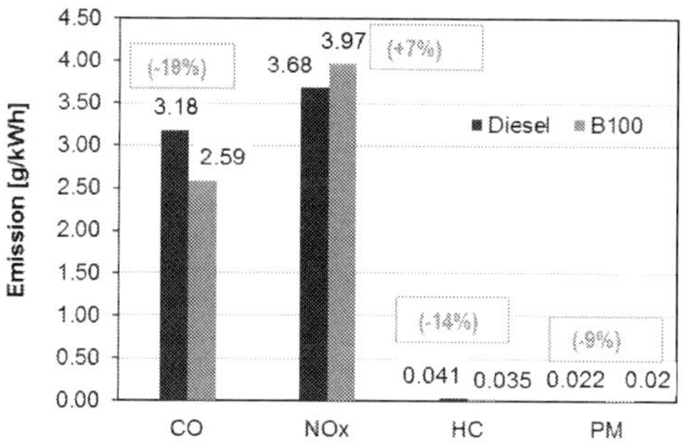

Figure 14. Comparison of the exhaust emissions measured during the road tests

5. Conclusions

The investigations performed confirm that the B100 fuel is an interesting alternative to diesel fuel. The assessment of the environmental impact of B100-fuelled vehicles is not unambiguous and the presented results do not fully confirm that the applied fuel is a more advantageous fuel in terms of exhaust emissions. The results obtained indicate a slightly lower emission of CO, HC and PM when the vehicle was fuelled with B100, but at the same time its application results in a slight increase in the emission of NO_x. It seems that similar level of exhaust emissions recorded regardless of fuel type (diesel fuel or RME) results from an advanced exhaust gas aftertreatment system which was applied in the test vehicle. It is noteworthy that the use of biofuels is advantageous in terms of CO_2 emission due to the phenomenon of closed cycle. Another important argument speaking for the use of biofuels is the preservation of natural resources of crude oil and the possibility of liberating from the crude oil supplies (at least to some extent) that in many European countries are imported consumer goods. It thus appears that the increased use of biofuels in transport is a good policy and the actions should be continued. Biofuels of the second generation (biofuels that are in agreement with food manufacturing policy) are beginning to receive increased attention. Their application does not raise controversy in terms of food shortages in the world, which in combination with environmental benefits becomes a perfect alternative to conventional fuels.

Author details

Jerzy Merkisz*, Paweł Fuć, Piotr Lijewski and Miłosław Kozak

*Address all correspondence to: Jerzy.Merkisz@put.poznan.pl

Poznan University of Technology, Institute of Combustion Engine and Transport, Poland

References

[1] Merkisz J., Kozak M. and Teodorczyk A. Korzyści i zagrożenia związane ze stosowaniem biopaliw w silnikach spalinowych (*Benefits and risks resulting from application of engine biofuels*). Archivum Combustionis. 2003;3(1)

[2] Lauperta M., Armas O. and Rodriguez-Fernandez J.. Effect of Biodiesel Fuels on Diesel Engine Emissions. Progress in Energy and Combustion Science. 2008;(34)

[3] Assessment and Standards Division, editors. A Comprehensive Analysis of Biodiesel Impacts on Exhaust Emissions. Office of Transportation and Air Quality of the US Environmental Protection Agency; 2002. DOI: EPA420-P-02-001

[4] Daisuke K., Hajime I. and Yuichi G. Effect of Biodiesel Blending on Emission Characteristics of Modern Diesel Engine. SAE. 2008; 2008-01-2384

[5] Durbin T.D., Cocker III D.R., Sawant A.A., Johnson K., Miller J.W., Holden B.B., Helgeson N.L. and Jack J.A.. Regulated Emissions from Biodiesel Fuels from On/Off-Road Applications. Atmospheric Environment. 2007;(47)

[6] Fontaras G., Karavalakis G., Kousoulidou M., Tzamkiozis T., Ntziachristos L., Bakeas E., Stournas S. and Samaras Z. Effects of Biodiesel on Passenger Car Fuel Consumption, Regulated and Non-Regulated Pollutant Emissions over Legislated and Real-World Driving Cycles. Fuel. 2009;(88)

[7] Fujia W., Jianxin W., Wenmiao C. and Shijin S. A Study on Emission Performance of a Diesel Engine Fueled with Five Typical Methyl Ester Biodiesels. Atmospheric Environment. 2009;(43)

[8] Graboski M.S. and McCormick R.L. Combustion of Fat and Vegetable Oil Derived Fuels in Diesel Engines. Prog. Energy Combust. Sci. 1998;24

[9] Lei Z., Wugao Z., Wei L. and Zhen H. Experimental Study on Particulate and NOx Emissions of a Diesel Engine Fueled with Ultra Low Sulfur Diesel, RME-Diesel Blends and PME-Diesel Blends. Science of the Total Environment. 2010;(408)

[10] Merkisz J., Kozak M., Pielecha J. and Andrzejewski M. The Influence of Application of Different Diesel Fuel-RME Blends on PM Emissions from a Diesel Engine. Combustion Engines. 2012;(1)

[11] Nigro F., Trielli M. and Costa C. Emission Characteristics of a Diesel Engine Operating with Biodiesel and Blends. SAE. 2007; 2007-01-2635.

[12] Szlachta Z. Zasilanie silników wysokoprężnych paliwami rzepakowymi (*The application of rapeseed fuels for compression ignition engines*). Warszawa: WKiŁ; 2002.

[13] International Association of Public Transport: Towards low/zero-carbon urban mobility in Europe - position paper. November 2011. http://www.uitp.org/sites/default/files/cck-focus-papers-files/FPNov2011.pdf.

[14] Documentation and Evaluation of International Case Studies Compiled within the European research project "Deriving effective least-cost policy strategies for alternative automotive concepts and alternative fuels – ALTER-MOTIVE", March 2010.

[15] Clean Buses – Experiences with Fuel and Technology Options. February 2014. http://www.clean-fleets.eu.

Chapter 3

The Study on the Influence of Diesel Fuel Oxygenates on Exhaust Emissions

Miłosław Kozak

Additional information is available at the end of the chapter

http://dx.doi.org/10.5772/62005

Abstract

The study discusses the use of oxygenated fuels in reduction of exhaust emissions from diesel engines. The study analyzes the physicochemical properties of oxygenated compounds in diesel fuels based on which 12 such compounds were selected for experimental research (glycol ethers, maleates, carbonates and alcohols). The study presents the results of investigations of the influence of oxygenated fuels on the exhaust emissions under dynamic homologation cycle chassis dynamometer conditions (diesel passenger vehicle, NEDC and FTP-75 cycles). The relationship among the content of individual oxygenated compounds, the conditions of the tests cycle and the exhaust emissions has been ascertained. The performed studies and experimental research have shown that the application of oxygenated fuels in diesel engines results in a significant reduction of the PM emission at a small increase in the emission of NO_x. Changes in the emission of CO and HC depend not only on the oxygen content in the fuel but also on the self-ignition quality of the applied oxygenated compounds. The application of oxygenated fuels does not influence the emission of CO_2. Out of the analyzed oxygenated compounds, the most advantageous emission changes (reduction of the emission of PM, CO and HC at a small increase in the emission of NO_x) are generated by the compounds of the glycol ethers group.

Keywords: Diesel engine, diesel fuel, oxygenates, exhaust emissions

1. Introduction

An efficient transport system is one of the basic elements of every developed economy in the world. Transport is also a very important aspect of life for a large part of humanity, because it facilitates the daily movement of people. Access to transport is one of the main factors affecting the quality of life. In addition to the major benefits of transportation, there are also the adverse consequences, particularly pollution.

Vehicles are a major source of air pollution in developed countries, including the European Union (EU) (Fig. 1a). It should be noted that in the past 20 years, the carbon monoxide (CO) and hydrocarbons (HC) emissions from transport have decreased significantly, while NO_x emissions, and in particular particulate matter (PM), have declined only slightly (Fig. 1b).

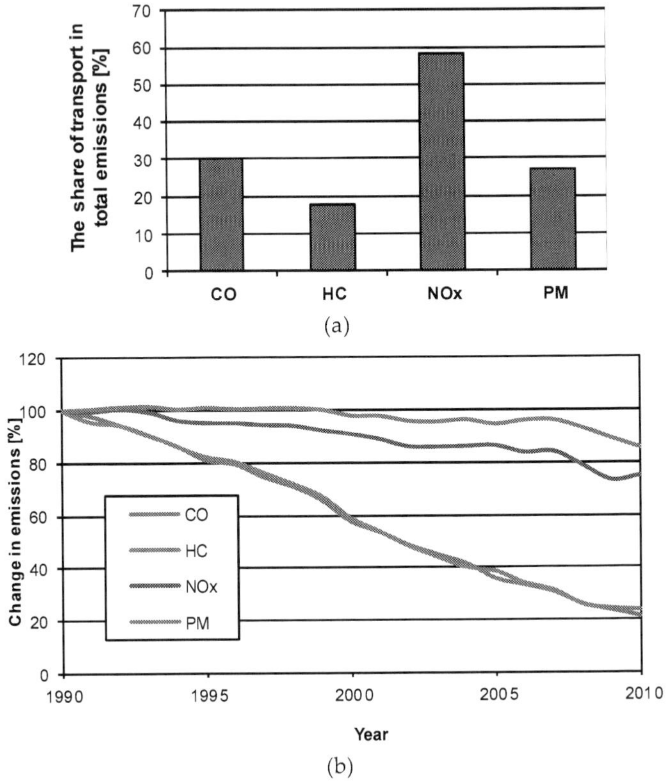

Figure 1. The share of transport in total emissions of major air pollutants in the European Union in 2009 (a) and change in emissions of these pollutants in the transport sector in 1990–2010 (b) [1]

The main sources of CO and HC emissions from transport are vehicles powered by gasoline engines. The share of vehicles with spark ignition (SI) and compression ignition (CI) engines in NO_x emissions is similar, while vehicles with diesel engines are responsible for the majority of PM emissions. Particulate matter is among the most harmful of emissions from motor vehicles. In the report of the US Environmental Protection Agency on the dangers of exhaust gases of diesel engines [2], it has been demonstrated, on the basis of several dozen independent medical studies on humans and laboratory tests on animals, that airborne PM have carcinogenic effects and contribute in particular to the occurrence of lung cancers.

One of the reasons of the mentioned small reduction in PM emissions compared to other exhaust compounds is an increase in the number of sources of PM emissions, which means the number of vehicles powered by diesel engines is higher. From the middle of the past decade, more cars with diesel engines than SI were sold in the European Union. In 2012, the share of passenger vehicles with diesel engines among new cars was 56% [3]. The economic development of the countries in the European Union leads to an increase in the number of heavy vehicles, which typically use only diesel engines. This situation is well illustrated by the trends in demand for motor fuel. In 2000, the European Union had a similar demand for automotive gasoline and diesel fuels. At present, the demand for diesel fuel is twice that of gasoline, and taking into account the off-road oil consumption, it is nearly three times higher (Fig. 2). A similar situation also exists in Poland, but in this country, it is largely the result of popularity of liquefied petroleum gas (LPG). According to data from the Central Statistical Office [4], 3 925 thousand tons of gasoline and 12 219 thousand tons of diesel fuel have been sold in Poland in 2011.

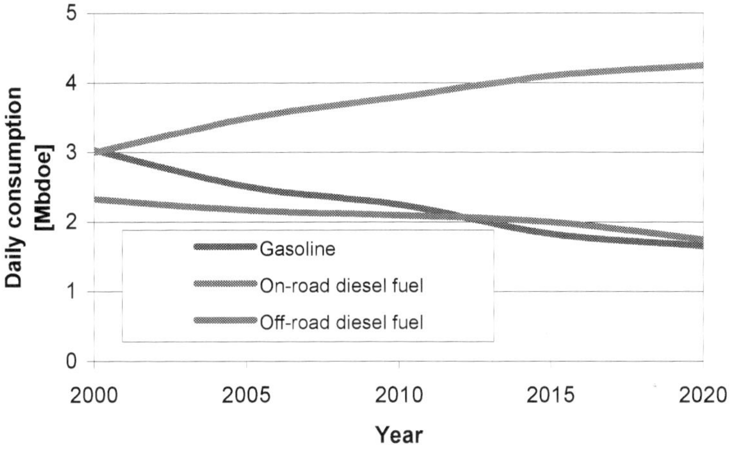

Figure 2. Daily consumption of motor fuels in the European Union [5]

The second reason for the similar level of PM emissions from transport sources continued over the years is lower PM emission reduction efficiency of vehicles compared to the gaseous components of the exhaust. Currently, getting acceptably low exhaust emissions from diesel engines requires numerous and varied measures, covering at least [6]:

- the optimization of the combustion process in the engine,
- utilization of systems that reduce the creation of harmful compounds in the operating processes of the engine (e.g., exhaust gas recirculation (EGR)),
- using effective exhaust aftertreatment systems,

- utilization of onboard diagnostics (OBD) to monitor the emission levels of harmful compounds by the vehicle,
- improving the fuel composition to reduce the creation of harmful exhaust compounds.

The basis for the development of eco-friendly motor fuels are research programs concerning the impact of fuel on emissions of toxic components, performance, durability, and other indicators for internal combustion engines. The first such program was AQIRP (Auto/Oil Air Quality Improvement Research Program) conducted in the US in 1989–1997. However, it only researched fuels for SI engines [7]. Evaluation of the impact of fuel emissions was also one of the elements of the JCAP program (Japan Clear Air Program) implemented in Japan in 1996–2001 [8]. In the years 1993–1996, the European Union conducted its EPEFE program (European Programme on Emissions, Fuels and Engine Technologies), usually abbreviated to Auto-Oil program [9]. On the basis of research [10], it was found that the greatest impact on the level of harmful exhaust emissions from diesel engines have the following parameters of diesel fuel (Fig. 3):

- cetane number,
- density,
- polycyclic aromatic hydrocarbons (PAH) content,
- fractional composition, particularly the amount of heavy hydrocarbons characterized by the T90, T95, and FBP temperatures,
- sulfur content.

Parameter	Unit	Since 2000	Since 2005
Cetane number, min.	–	51	51
Sulfur content, max.	ppm	350	50/10[1)]
PAH content, max.	% (m/m)	11	11
Density at 15°C, max.	kg/m^3	845	845
Distillation, T95 max.	°C	360	360

[1)] 10 ppm since 2009

Table 1. Specifications of diesel fuel according to the European Directive 98/70/EC

Auto-Oil program was the only program so far that so comprehensively encompassed the issues of the impact of fuel emissions of harmful exhaust compounds. The results of the Auto-Oil program were the basis for establishing the fuel specifications of the European Union, in force since 2000, contained in the European Directive 98/70/EC. The new requirements introduced by this directive significantly changed the parameters of diesel fuels affecting the level of exhaust emissions (Table 1). Through this, the potential of conventional diesel fuels in

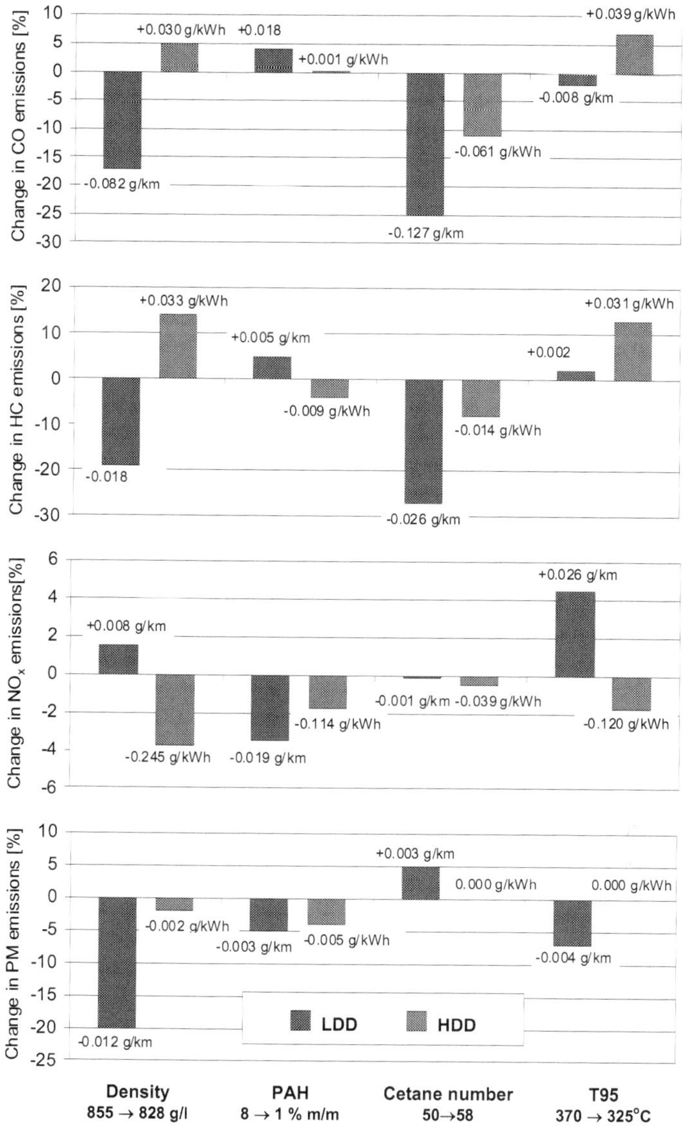

Figure 3. Effect of diesel fuel parameters on the exhaust emissions of the European LDD and HDD engines according to the Auto-Oil program [10]

reducing emissions was largely used up, so that further modification of the conventional fuel parameters is possible only to a limited extent. The sulfur content in fuels produced today is nearly zero, and the PAH content has been reduced to about 2–3%. Excessive increase in cetane number would have an adverse impact on the process of combustion and emissions. A further change of composition in the direction of lighter fractions is also unjustified. So to even further exploit the potential of fuels to reduce exhaust emissions of diesel engines, unconventional modifications of their properties are necessary. It seems that one of the promising fields of this type of activity is the use of oxygenated fuels, namely the introduction of components that act as carriers of oxygen (oxygenates) into the composition of diesel fuels. Oxygenated fuels are successfully utilized in SI engines to provide lower emissions of CO and HC. Due to the nature of combustion in the diesel engine, namely an inhomogeneous air–fuel mixture and local oxygen deficiency regions, the emission benefits associated with the use of oxygenated fuels in diesel engines should be even greater. This chapter is dedicated to the issue of the impact of oxygenated fuels on exhaust emissions from diesel engines.

It should be noted that the introduction of low-emission diesel fuels involves a number of potential benefits, which cannot be provided by other solutions for reducing emissions. First of all, eco-friendly fuels immediately impact the whole fleet of vehicles, and their introduction into the market in sufficient quantities (as opposed to alternative fuels) can be done in a short time. Although the unit emission reduction is probably less than one achieved with modern engine technology or alternative fuels, because of the huge scale of fuel consumption, a quick global effect may be larger than what can be achieved by several other means. In addition, low-emission fuels can provide a reduction in emissions from older vehicles, which on the one hand are the largest source of harmful exhaust emissions, and on the other hand, reducing their harmfulness is technically impossible or not cost-effective through other means. Another advantage of the low-emission diesel fuels is their impact on reducing harmful exhaust emissions in circumstances when other measures are not effective (e.g. engine start-up and warm-up period).

2. Requirements for oxygenated compounds used as components of diesel fuels

Oxygen compounds used as components of oxygenated fuels must demonstrate properties that are required for all components of motor fuels. Not all oxygenates are suitable for use as an additive to diesel fuel. According to Murphy [11], potential oxygenated components must meet the following requirements:

- be able to mix with diesel fuel in the whole range of engine operating temperatures,
- low-temperature parameters of diesel fuel and oxygenated compounds mixtures must be suitable for the climate where they will be used,
- mixtures of diesel fuel and oxygenated compounds should not be susceptible to stratification as a result of contamination with water,
- oxygenated compounds should have a sufficiently high cetane number,

- oxygenated compounds should not have a too high volatility nor too low flash point.

Additional important requirements for potential oxygenates of diesel fuels should include:

- high oxygen content in the molecule allowing for small concentrations of oxygenated compounds to achieve sufficient amount of oxygen in the final fuel mixture; lower concentrations also have a lesser impact on the physical properties of fuel,
- no negative impact on fuel system components: no corrosive effect on metals or swelling effects on plastics (elastomers),
- low own toxicity of the oxygenated compound,
- competitive price and availability.

It should be noted that the fulfillment of these criteria will often depend on the amount of oxygenated compound added to the fuel composition.

The literature provides various proposals for chemical compounds to be used as oxygen compounds in fuels. Particularly, one paper [12] presents an analysis of the physicochemical properties of 71 potential oxygenated compounds. After initial selection on the basis of three parameters, such as oxygen content, density, and flashpoint temperature, 42 oxygenated compounds were addressed for further analysis, where they have been evaluated in terms of miscibility with diesel fuel and water tolerance. Seventeen compounds that at this stage rated positively were then tested for corrosive properties. Their toxicological properties, biodegradability, and impact on elastomers were also evaluated. Based on the analyses performed, eight most promising oxygenated compounds were chosen for further use in diesel fuels: tripropylene glycol monomethyl ether, 1-methoxy-2-propanol, dipropylene glycol monomethyl ether, 2-ethoxy ethyl acetate, 2-ethoxy ethyl ether, diethyl adipate, tributyrin, and dibutyl maleate.

Bertoli and Boulouchos [13] describe the analysis of 27 prospective oxygen compounds in terms of their compatibility with diesel fuel, their oxygen content, and calorific value. This analysis found that the best oxygenated compounds are diethylene glycol dimethyl ether, diethylene glycol diethyl ether, dibutoxymethane (butylal), and dipenthylether. Delfort et al. [14] show the applicability as fuel components of 18 oxygen compounds of chemical groups such as ethers, acetals, polyacetals, and carbonates. Murphy in his works [15, 16] pointed out the inappropriate fire safety and toxicology performance of some of oxygen compounds.

3. Analysis of exhaust emissions from a passenger car fueled with oxygenated fuels

3.1. The essence of application of oxygenated diesel fuels

Fuel, as the integral element of combustion in the engine, has an immense effect on the composition and harmfulness of exhaust gases emitted by the engine. The changes of EU standard requirements for diesel fuels, implemented in 2000, 2005, and 2009, were directed toward the reduction of the harmfulness of the exhaust gases emitted by the engines. At present, the modification of the conventional properties of diesel fuel that affect the emissions (cetane number, sulfur content, distillation, etc.) is possible to a limited extent only. Therefore, in order to take the maximum advantage of the fuel potential in exhaust emissions reduction, some more unconventional modifications of the fuel should be considered. The application of oxygenated compounds seems to be one of the more promising solutions in this field. The compounds particularly affect a very troublesome component of diesel engine exhaust – i.e., particulate matter. On the basis of physicochemical and environmental parameters, test results described in the literature, as well as price and availability, oxygen compounds belonging to chemical groups such as glycol ethers, maleates, and carbonates were selected for further experimental testing in this study. The study also includes butanol, whose mass production from biomass is being considered. Butanol is closer in terms of its properties to diesel fuel than the currently produced large-scale ethanol. The detailed characteristics of the oxygen compounds selected for testing are presented in Section 3.2.

The literature presents some test results of maleates as components of diesel fuel. Among them, there are works delivered by Stoner and Lizinger [17], who obtained reduction in PM and NO_x emissions at the same time by using diethyl maleate and dibutyl maleate. They also tested some glycol ethers, which, however, caused less favorable changes in emissions than maleates. Natarajan et al. [12] analyzed potential oxygenated compounds and, based on their physical and chemical properties, they selected eight of the most suitable oxygenates to be used in diesel fuel. Among them, there was dibutyl maleate. Further, they performed emissions tests in one mode of engine operation with the use of the selected components at the concentration, ensuring 7% of oxygen in the fuel [18]. Dibutyl maleate caused considerable decrease in PM emissions and, at the same time, considerable increase in NO_x emissions. Based on total PM emissions and other considerations, dibutyl maleate and tripropylene glycol monomethyl ether were identified as the most promising oxygenated compounds for future engine testing. These more extensive engine tests carried out in several engine operating modes confirmed high efficiency of the abovementioned oxygenates in PM emissions reduction and showed that these oxygenates had no unfavorable influence on NO_x emissions [19].

In the available bibliography, there are some examples of the results obtained in the tests of carbonates as diesel fuel components. Murayama et al. [20] carried out some extensive research on the influence of dimethyl carbonate on the combustion characteristics and emissions from a direct injection (DI) diesel engine. It was found that for the carbonate content in fuel of 10%, the smoke emission was lowered by 35–50%, and also that the extent of the smoke emission reduction was proportional to the oxygenate content. Moreover, it was found that the carbonate additive caused a significant decrease in the CO and HC concentrations in exhaust gases and a small increase in the NO_x emissions. Also Akasaka and Sakurai presented in their paper [21] the reduction in smoke emission obtained with the use of an additive from the carbonate group (its name was not mentioned). The research on the influence of oxygenates on the PM emissions in the New European Driving Cycle (NEDC) with the use of the Audi A3 TDI (Euro 2) passenger car were carried out by Delfort et al. [14]. They tested 18 oxygenates, including 5 oxygenates from the chemical family of carbonates. They found that both oxygen content in fuel and the oxygenated compound chemical structure were decisive for the reduction in the PM emissions. Their tests showed that at the same content of different oxygenates in fuel, the highest reduction in the PM emissions was obtained in case of the use of carbonates. Next, from among all tested carbonates, diethyl carbonate and diethoxy-2-ethyl carbonate were the most effective ones, which, when added to the fuel in amount of 5%, caused the reduction in the PM emissions in the NEDC by 17–18%. Substantial reduction in exhaust soot emissions for diesel fuel containing dimethyl carbonate was described by Kocis et al. [22]. They also discussed the potential paths of oxygenated compounds influence on soot production.

Glycol ethers are as often as not tested as diesel oxygenated compounds. The aforementioned research, carried out by Delfort et al. [14], showed their moderate efficiency in PM emission reduction. The research conducted by Hallgren and Heywood [23] also showed that glycol ethers decrease PM emissions. Moreover, this research proved that these oxygenates reduce mainly insoluble fraction (soot), whereas soluble organic fraction (SOF) remains intact. Yeh et al. [24] proved that glycol ethers cause not only a reduction in PM emissions but also a reduction in CO and HC emissions and an increase in NO_x emissions. Favorable effects of glycol ethers on exhaust emissions were also described in papers delivered by Porai et al. [25] and Akasaka and Sakurai [21].

The majority of the published papers concerning the influence of oxygenates on exhaust emissions from diesel engines have reported the results of research conducted under stationary conditions, for one or more engine operating modes. From among the few investigations carried out under transient conditions, aforementioned research performed by Delfort et al. [14] and Yeh et al. [24] should be noted.

The detailed mechanism by which oxygenates reduce PM emissions has not been precisely defined so far. The literature contains contradictory results as to whether the oxygen content in fuel is the only important factor determining PM/smoke emissions, or whether oxygenate molecular structure or other variables also play important roles. Research carried out by Nabi et al. [26] with highly oxygenated diesel fuels containing glycol ethers and carbonates showed that smoke decreased linearly and depended entirely on the oxygen content in the fuel, regardless of the kind of the oxygenate. Another study with highly oxygenated diesel fuels by Mueller and Martin [27] indicates that overall oxygen content is not the only important parameter governing the soot reduction potential of a fuel, and that molecular structure effects may also be significant. Buchholz et al. [28] using selective labeling of carbon atoms in dibutyl maleate molecules proved that the molecular structure and distribution of oxygen within an oxygenated diesel fuel strongly influences the ability of the fuel to reduce PM emissions. In the recent study written by Boot et al. [29], it was shown that the oxygen content has a pronounced effect on PM reduction, but chemical structure effects can be almost of the same order.

The experimental results presented in the chapter were obtained within a research program investigating the effect of different synthetic oxygenates on exhaust emissions from diesel passenger cars. The objective of this study was to select the most promising oxygenate compounds as blending components in diesel fuel for further advanced testing and practical application. The research was carried out with oxygenates representing different chemical groups. All the emissions test were conducted over dynamic cycles in a chassis dynamometer (NEDC, FTP-75).

3.2. Research methodology

3.2.1. Scope of the research

The study was performed at the BOSMAL Automotive Research & Development Institute in Bielsko-Biała. They included measurements of toxic exhaust emissions (CO, HC, NO_x, and PM), CO_2 emissions, and fuel consumption on a chassis dynamometer in test cycles for passenger cars with a Euro 4 class diesel engine run on different oxygenated fuel variants. The conducted study was comparative; hence, measurements were also made for cars running on conventional diesel fuel, not containing oxygenated components. The research was staged and includes, in particular:

1. Preliminary studies (NEDC) aimed to determine the effect of each of the 12 researched oxygenated compounds on the exhaust emissions, thus indicating which of the tested fuel additives are most beneficial for reducing the engine exhaust toxicity. The results of the preliminary research were also used as the basis for determining the unit efficiency of oxygen contained in the fuel to reduce PM emissions as well as for the selection of oxygen compounds for further, more detailed analysis.

2. Investigation of the effect of oxygenates on exhaust emissions, depending on their content in the fuel (5 or 10% v/v, NEDC).

3. Investigation of the effect of oxygenates on exhaust emissions, depending on the test cycle conditions (NEDC and FTP-75).

3.2.2. Research fuels

The study used 12 different synthetic oxygenates representing four chemical groups, such as the glycol ethers, maleates, carbonates, and alcohols. Oxygenated compounds were used in concentrations of 5 and 10%. Selection of components was dictated primarily by their physicochemical properties (oxygen content, compatibility with diesel fuel, etc.) and the results of tests described in the literature, followed by price and availability. Physicochemical properties of the oxygen compounds used in the study are shown in Table 2. The oxygen content of the compounds used is in the range 21.6–53.3% m/m. Oxygen compounds were labeled in this work with the symbol PT-X-Y%, where X is the number of oxygen compound used in accordance with Table 2, and Y is its content by volume.

All research fuels were formulated based on Euro 5 grade diesel fuel. Due to the long duration of the research, and taking into account the aging processes of fuel (change in fuel properties during storage), diesel fuel used was purchased in two tranches, labeled in this work as ON I and ON II. The properties of diesel fuel in both tranches were similar; they have been shown in Tables 3 and 4.

No.	1	2	3	4	5	6
Chemical name	Ethylene glycol dimethyl ether	Diethylene glycol dimethyl ether	Triethylene glycol dimethyl ether	Tetraethylene glycol dimethyl ether	Diethylene glycol dibutyl ether	Dipropylene glycol dimethyl ether
Molecular formula	$CH_3OCH_2CH_2OCH_3$	$(CH_3OCH_2CH_2)_2O$	$CH_3O(CH_2CH_2O)_3CH_3$	$CH_3O(CH_2CH_2O)_4CH_3$	$(CH_3CH_2CH_2CH_2OCH_2CH_2)_2O$	$CH_3OC_3H_6OC_3H_6OCH_3$
Molecular weight, amu	90.12	134.17	178.23	222.28	218.33	162.23
Oxygen content, % m/m	35.6	35.8	36.0	36.0	22.0	29.6
Boiling point, °C	85	162	220	275	256	175
Freezing point, °C	–69	–64	–40	–30	–60	–80
Flash point, °C	–6	51	113	141	120	65
Density @ 20°C, kg/m³	867	944	987	1010	884	903
Viscosity @ 20°C, mm²/s	0.5	1.2	2.5	4.1	2.3	1.12
Cetane number (calculated)	86	112	144	144	144	86
Vapor pressure @ 20°C, kPa	6.4	0.4	–	<0.001	<0.001	–
No.	7	8	9	10	11	12
Chemical name	Tripropylene glycol methyl ether	Diethyl maleate	Dibutyl maleate	Dimethyl carbonate	Diethyl carbonate	N-butanol
Molecular formula	$CH_3(OC_3H_6)_3OH$	$C_2H_5OCOCH=CHCOOC_2H_5$	$CH_3(CH_2)_3OCOCH=CHCOO(CH_2)_3CH_3$	$(CH_3O)_2CO$	$(C_2H_5O)_2CO$	$CH_3(CH_2)_3OH$
Molecular weight, amu	206.28	172.18	228.28	90.08	118.13	74.12
Oxygen content,	31.1	37.3	28.1	53.3	40.7	21.6

% m/m						
Boiling point, °C	243	225	281	90	127	117
Freezing point, °C	−78	−10	–	3	−43	−90
Flash point, °C	121	–	>110	–	–	34
Self-ignition temperature, °C	227	350	–	–	–	343
Density @ 20°C, kg/m³	963	1064	988	1069	975	810
Viscosity @ 20°C, mm²/s	5.5	–	–	0.56	0.78	3.64
Cetane number (calculated)	58.3	56.8	43.8	10.3	13.8	10.2
Vapor pressure @ 20°C, kPa	<0.001	0.1 (14°C)	–	2.4	1.3	0.5

Table 2. The properties of synthetic oxygenated compounds used in this research

		Unit	Value
Cetane number		–	51.8
Cetane index		–	53.1
Density @ 20°C		kg/m³	831.2
Sulfur content		ppm	9.9
Oxygen content		% m/m	0.0
Viscosity	@ 20°C	mm²/s	4.283
	@ 40°C	mm²/s	2.773
Distillation	E250	% v/v	35.3
	E350	% v/v	97.6
	T95	C	341.6
	FBP	C	352.0
Aromatic hydrocarbons	Total aromatics	% m/m	23.9
	Monoaromatics	% m/m	21.4

		Unit	Value
	Diaromatics	% m/m	2.4
	Tri + aromatics	% m/m	0.1
	Total PAH	% m/m	2.5

Table 3. Specifications of the first batch of base diesel fuel (ON I)

		Unit	Value
Cetane number		–	52.8
Cetane index		–	53.4
Density @ 20°C		kg/m^3	827.7
Sulfur content		ppm	8.8
Oxygen content		% m/m	0.0
Viscosity	@ 20°C	mm^2/s	4.096
	@ 40°C	mm^2/s	2.607
Distillation	E250	% v/v	38.1
	E350	% v/v	–
	T95	C	332.3
	FBP	C	343.7
Aromatic hydrocarbons	Total aromatics	% m/m	20.7
	Monoaromatics	% m/m	18.8
	Diaromatics	% m/m	1.7
	Tri+ aromatics	% m/m	0.2
	Total PAH	% m/m	1.9

Table 4. Specifications of the second batch of base diesel fuel (ON II)

3.2.3. *Research conditions*

3.2.3.1. *Research equipment description*

The research was conducted in the Exhaust Toxicity Laboratory at the BOSMAL Automotive Research & Development Institute in Bielsko-Biała, which has the instrumentation to measure toxic exhaust emissions from vehicles powered by diesel engines and accredited by the TÜV. The course of research in the laboratory is controlled by the measurement management system AVL CESAR. The temperature and humidity at the level required by the testing procedure are maintained by a complete heating, ventilation, and air-conditioning (HVAC) system.

The laboratory is equipped with a SCHENCK 500GS/60 chassis dynamometer (Fig. 4). Chassis dynamometer is a device designed to simulate the resistance of motion (rolling and aerodynamic) for a car moving forward on a flat road and the inertia of the vehicle.

The core elements of the dynamometer include:

- A set of rollers with individual ball bearings, combined with a system of clutches and a belt drive, powered by a DC motor,
- DC motor fitted with a thyristor control system,
- A set of rotating weights, used for mapping the inertial mass of the tested vehicle,
- A set of controls and measurement devices; the entire operation of the dynamometer (e.g. setting the inertial mass, characterization, calibration, etc.) is done via its own test bench control computer.

Dynamometer is equipped with a parameter adjustment system for the driver (Driver Guiding Unit LMC 68/40). The operation of this device consists in presenting (in advance) the required driving performance parameters as a function of time to the driver: speed, moving and stopping points for the vehicle, as well as the time to push the clutch down and shifting gears. In addition, the current time of the test as well as the number and total time of breaching the preset parameters of the drive cycle are given.

Figure 4. Chassis dynamometer SCHENCK 500GS/60 in BOSMAL Automotive Research & Development Institute

The measurement of toxic exhaust emissions during the driving tests was carried out with an exhaust collection and sampling system with a Venturi nozzle – "Quadro Venturi" (CFV-CVS)

http://dx.doi.org/10.5772/62005

of the AVL CEC-Q20 type, equipped with a full-flow exhaust dilution tunnel AVL CET-LD/20. A sample of the mixture of exhaust and air, in proportion to the flow rate of the mixture, is continuously taken and collected in a sampling bag during the driving test.

After completing the test, the contents of the measurement bags, which is a mixture of exhaust gases and ambient air, is directed to a set of analyzers, which determine the concentration of each of the gaseous exhaust components. In accordance with the procedure of measuring emissions from vehicles equipped with CI engines, the exception is the measurement of the concentration of HC, which is done by drawing a sample through a heated line directly from the dilution tunnel. The sample for measuring exhaust emissions of PM is also taken from the dilution tunnel using a particulate sampling system AVL CEP-LD/100. The concentration measurements of gaseous exhaust components was performed through a set of AVL CEB 600 exhaust gas analyzers, which includes the Rosemount NGA 2000 device.

3.2.3.2. *Drive cycle: NEDC*

Most measurements of exhaust emissions are based on NEDC, introduced in the European Union on January 1, 2000, as a vehicle certification test. This cycle (Fig. 5) consists of two parts or phases (subtests): the UDC (Urban Driving Cycle) and the EUDC (Extra Urban Driving Cycle).

During the tests, the ambient temperature was 22–23°C, where the allowed temperature range for the test procedure is between 20 and 30°C. Before each test, the vehicle was conditioned in a laboratory environment. One test was performed per day. After each change of the fuel, the car also passed a whole cycle without the measurement of emissions to remove any residue of the previous fuel from the supply system, and the proper test was carried out the next day after reconditioning.

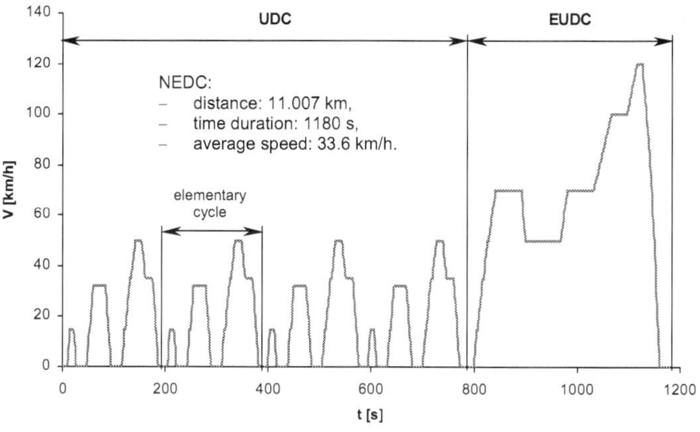

Figure 5. NEDC drive cycle according to the 98/69/EC directive

The intake of the analyzed exhaust gas starts immediately after starting the engine of the vehicle. The first phase (UDC) is composed of four elementary tests repeated in succession. Due to prior conditioning, the engine and the vehicle at the beginning of the UDC phase are at the ambient temperature, thus the start of the engine is cold, just as in real life at start-up after a long (e.g. overnight) parking. During the engine cycle, the temperature of the engine oil, engine coolant, and exhaust gases increases. The EUDC phase starts immediately after the end of the UDC phase, and therefore during the EUDC, the vehicle engine has warmed to operating temperature. In this phase, there are longer periods of operation of the engine at a constant speed and the vehicle speed reaches a value much higher than that in the UDC.

3.2.3.3. Drive cycle: FTP-75

The American federal test FTP-75 (Federal Test Procedure) is used in the US for the measurement of exhaust emissions by light-duty type vehicles and has been in force since 1996. This test consists of three phases (Fig. 6):

- cold start phase,
- transient phase,
- hot start phase.

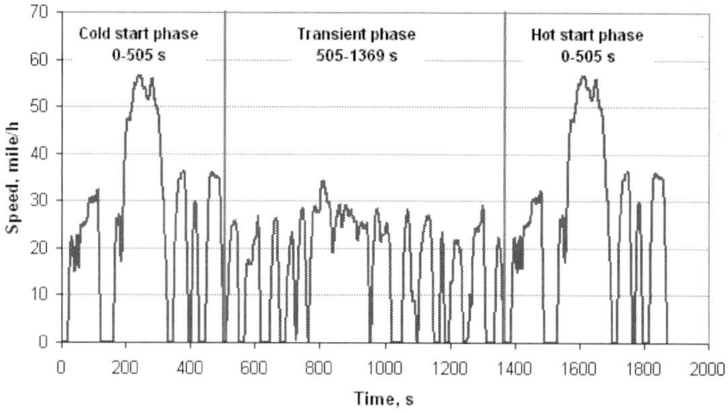

Figure 6. The result of an American FTP-75 drive cycle

A ten-minute break is made between the phases 2 and 3, during which the engine is switched off. The characteristics of the FTP-75 test are:

- distance traversed: 17.77 km,
- time duration: 1874 s,
- average speed: 43.1 km/h.

3.2.3.4. *Test vehicle*

The study used a Fiat Panda 1.3 16v Multijet vehicle, which at the start of the research represented the level of technological advancement of vehicle technologies. The car was equipped with a turbocharged diesel engine fueled in a common rail system. The car had controlled EGR system and catalytic exhaust aftertreatment, and its exhaust emissions met the Euro 4 standard. Key parameters of the test vehicle are presented in Table 5.

Due to the necessity of frequent changes in the fuel type for the test vehicle, and to avoid contamination of the test fuel with the previous fuel remaining in the fuel system, the fuel supply system of the vehicle was modified. A portable external fuel tank with its own electric low-pressure pump was used, delivering fuel directly into the high-pressure pump nozzle. Thanks to this, the length of the low-pressure fuel lines (i.e. the distance between low-pressure pump and the high-pressure pump) has been reduced several times. Several elements with a large capacity (the factory-issued fuel tank, low-pressure pump and fuel lines leading to the high-pressure pump), which could retain the fuel of the preceding tests, have been eliminated from the fuel supply system.

Vehicle type	Passenger car
Dry weight	950 kg
Engine type	Diesel, 4-cylinder in-line
Displacement	1.3 dm^3
Max. power	51 kW @ 4000 rpm
Max. torque	145 Nm @ 1500 rpm
Injection/combustion type	Direct injection common rail, turbocharged (intercooled)
Exhaust gas recirculation	Electronically controlled (closed-loop)
Emission control	Oxidation catalyst
Calibrated to	EURO 4

Table 5. Specifications of the test vehicle

3.3. Preliminary studies of the effect of oxygenated compounds on exhaust emissions

The aim of the research was to conduct a preliminary selection of oxygenates as fuel additives to ensure the most favorable changes in the composition of exhaust gases. These studies used 12 selected oxygenated compounds belonging to the following chemical groups: glycol ethers, maleates, carbonates, and alcohols. The preliminary studies testing the effectiveness of selected fuel oxygenates have been made under NEDC conditions, with a separate measurement of emission components in both its phases (UDC and EUDC). The emissions of CO, HC, NO_x, PM, and CO_2 were measured. The oxygenated fuels used for the study were blended using two base diesel fuels: ON I and ON II. PT-1 to PT-6 fuels contained the ON I diesel fuel and the others, i.e. PT-7 to PT-12 the ON II diesel fuel. Vehicle emission test results of the researched individual oxygenated fuels should therefore be referred to the results obtained for the corresponding base diesel fuels.

3.3.1. *Carbon monoxide emissions*

CO emission in the UDC phase (Fig. 7) was reduced when using most of the fuels containing glycol ethers, in particular for fuels containing oxygenated components 3–5. Note that these components had a very high cetane number and physical properties similar to those of diesel fuel. It appears that in the UDC phase conditions, so during engine warm-up, increased ability to self-ignition of fuel is essential to CO emission reduction. This is indicated by the result obtained for a fuel containing component 5 with a distinctly lower oxygen content than the components 3 and 4, as well as the results of studies on the effect of cetane number on emissions of CO made using fuels containing no oxygen [30, 31]. For fuels containing maleates and carbonates, the obtained CO emission levels were higher than for pure diesel. Maleates were characterized by a moderate amount of oxygen with a similar cetane number to that of diesel fuel, and for carbonates, low cetane number was compensated by a large oxygen content. Higher CO emissions have also been observed for the fuel containing *n*-butanol, which is a component with a low cetane number, relatively low oxygen content, and much lower boiling temperature than diesel fuel.

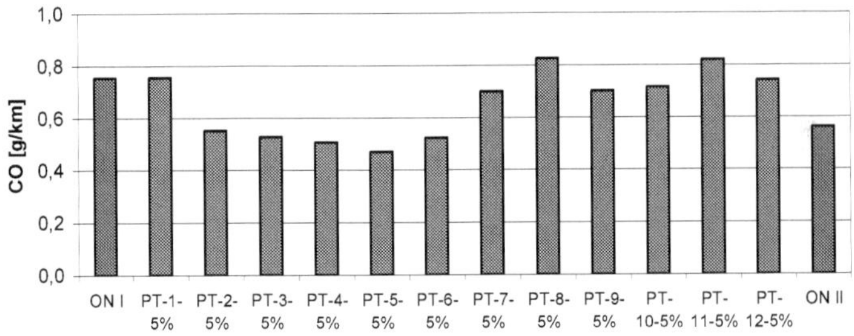

Figure 7. CO emissions during the UDC for conventional diesel fuels (ON I and ON II) and oxygenated fuels containing individual oxygenates at a concentration of 5% v/v; PT-1-5% to PT-6-5% were formulated on the basis of diesel fuel ON I and PT-7-5% to PT-12-5% were based on the ON II fuel

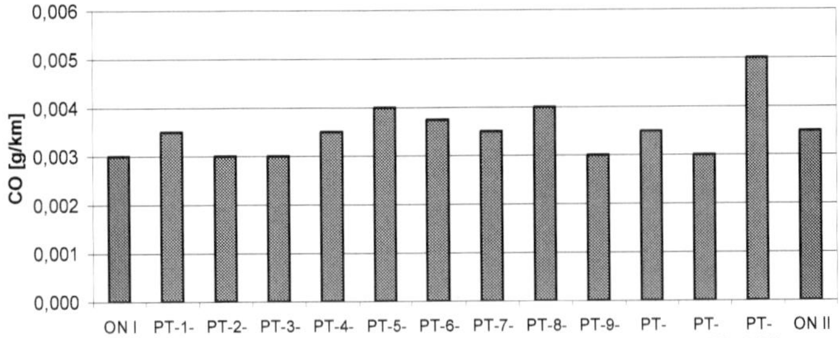

Figure 8. CO emissions during the EUDC for conventional diesel fuels and oxygenated fuels containing individual oxygenates at a concentration of 5% v/v

In the EUDC phase, the CO emissions are reduced to very low values (Fig. 8), mainly due to a stabilized thermal state of the engine and the effective operation of the catalytic converter. All fuels, except PT-12-5%, showed a similar level of emissions.

Very large disparities in CO emissions between phases UDC and EUDC (a difference of more than two orders of magnitude) mean that the average CO emissions in the NEDC (Fig. 9) are determined by the UDC phase. Therefore, the best performance over the NEDC was reported for the same fuel as for the UDC phase, thus containing glycol ethers.

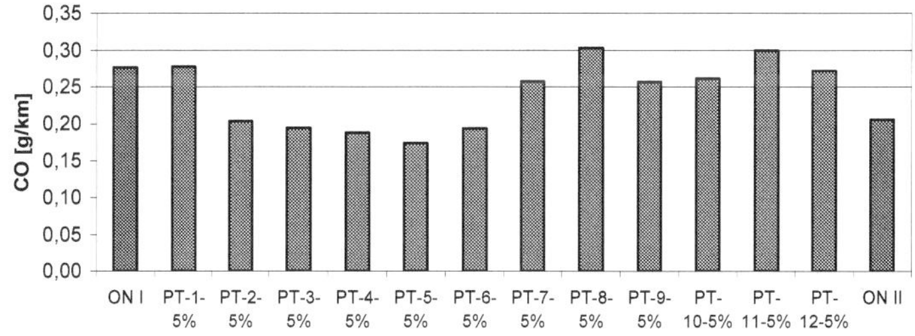

Figure 9. The average emissions of CO in the NEDC (UDC + EUDC) for conventional diesel fuels and oxygenated fuels containing individual oxygenates at a concentration of 5% v/v

3.3.2. *Hydrocarbons emissions*

Same as for CO, the UDC phase saw a significantly lower HC emission for most of the oxygenated fuels (Fig. 10). Again, the relationship of fuel cetane number and the levels of emission reduction has been noted. When using fuels containing maleates, carbonates, and *n*-butanol, a slightly higher emission of HC has been observed than for neat diesel fuel.

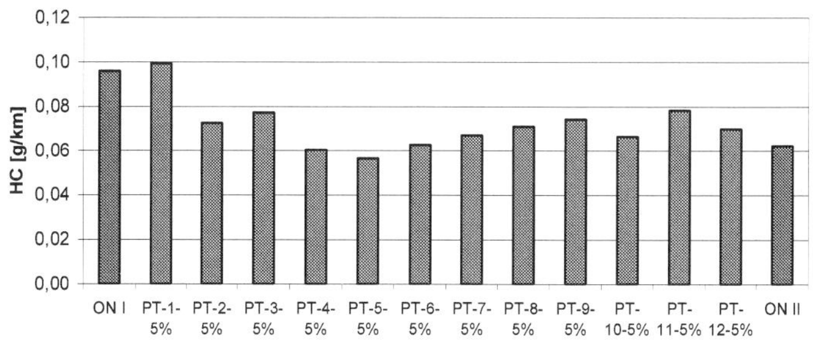

Figure 10. HC emissions during the UDC for conventional diesel fuels and oxygenated fuels containing individual oxygenates at a concentration of 5% v/v

HC emissions in the EUDC phase are roughly an order of magnitude lower than in the UDC phase (Fig. 11), which is the result of steady thermal state of the engine, the effective operation of the catalytic converter and less dynamic changes in the operating conditions of the engine. For all oxygenated fuels, the HC emissions were reduced, or at least maintained at the level similar to that of diesel fuel (PT-1-5%, PT-9-5%, PT-11-5%).

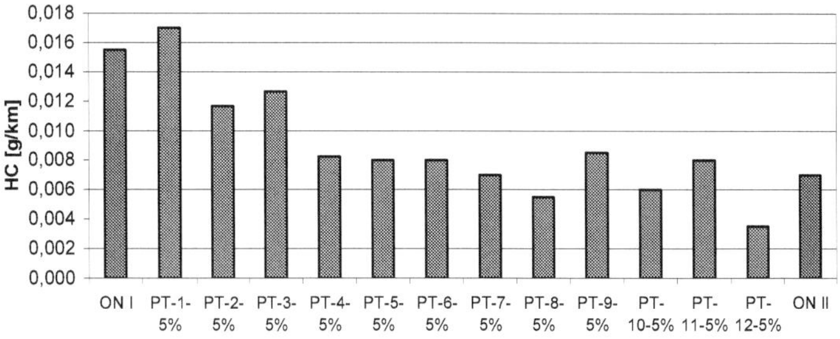

Figure 11. CO emissions during the EUDC for conventional diesel fuels and oxygenated fuels containing individual oxygenates at a concentration of 5% v/v

The average HC emission in the NEDC test (Fig. 12) is determined mainly by the UDC phase. Generally, it can be said that all tested glycol ethers, with the exception of the compound number 1, produced a significant (approximately 30%) reduction in HC emissions. The said ether has a lower cetane number and physical properties (viscosity, boiling point) greatly deviating from the characteristics of diesel fuel. In contrast, other fuels containing oxygenated compounds cause HC level to remain similar to or slightly higher than that of diesel fuel (fuels containing components 9 and 11).

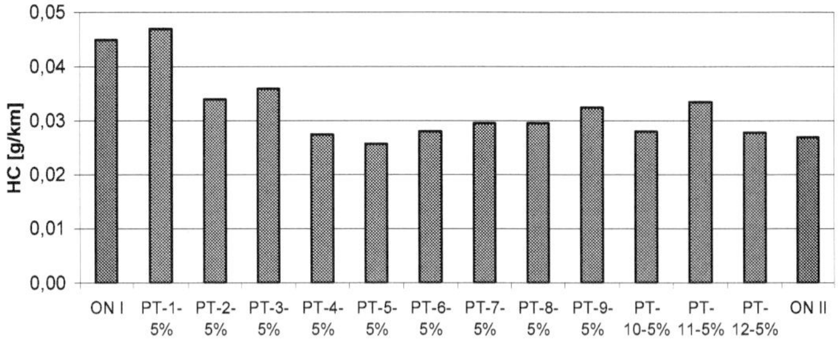

Figure 12. The average emissions of HC in the NEDC (UDC + EUDC) for conventional diesel fuels and oxygenated fuels containing individual oxygenates at a concentration of 5% v/v

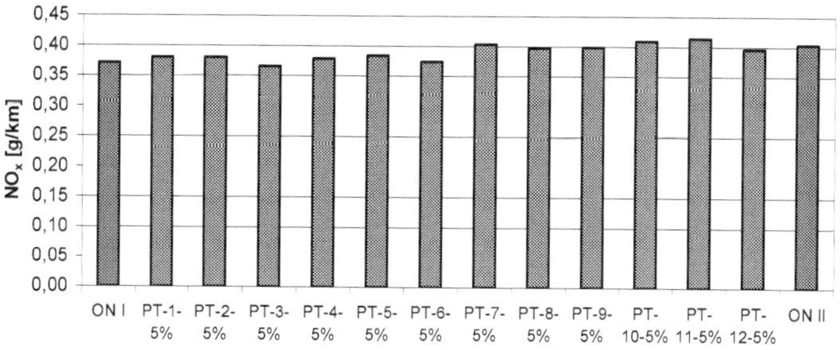

Figure 13. NO_x emissions during the UDC for conventional diesel fuels and oxygenated fuels containing individual oxygenates at a concentration of 5% v/v

3.3.3. Nitrogen oxides emissions

In the UDC phase, a very slight difference, not exceeding a few percent, was found in NO_x emissions for various fuels (Fig. 13). In the EUDC phase, for all the tested oxygenated fuels slightly higher NO_x emissions than for diesel fuel were observed (Fig. 14), but these differences can be considered minor.

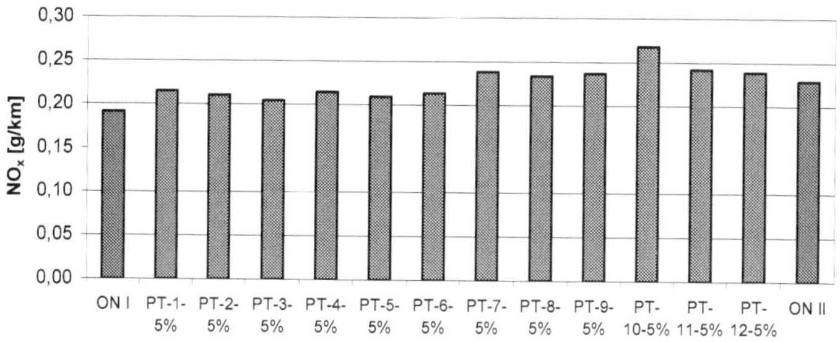

Figure 14. NO_x emissions during the EUDC for conventional diesel fuels and oxygenated fuels containing individual oxygenates at a concentration of 5% v/v

The average NO_x emission in the NEDC for most oxygenated fuels was slightly (only a few percent) higher than the recorded emissions when powered by diesel fuel alone (Fig. 15); for other oxygenated fuels, the emission values remained in line with diesel fuel. These results confirm the limited impact of fuel on the engine NO_x emission level (downstream the outlet valve). It should be added that the highest NO_x emissions were observed for fuel with the highest oxygen content, which may be explained by a better availability of oxygen in the zone where nitrogen oxides are created according to the *prompt* mechanism as well as by locally leaner air–fuel mixture, allowing for burning more fuel and achieving higher local temperatures in the combustion chamber. One can assume that the smaller intensity of the formation of NO_x was the effect of a high cetane number of fuels containing the components 1–6. This was confirmed in studies using fuels containing no oxygen [31].

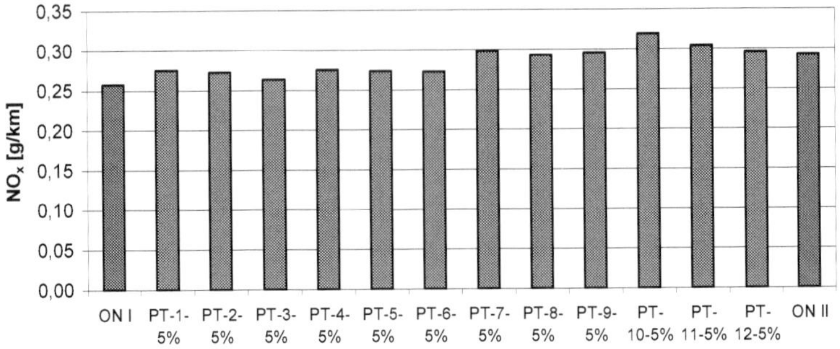

Figure 15. The average NO_x emissions in the NEDC (UDC + EUDC) for conventional diesel fuels and oxygenated fuels containing individual oxygenates at a concentration of 5% v/v

3.3.4. Particulate matter emissions

PM emissions in the UDC phase was lower for all oxygenated fuels (Fig. 16), except PT-12-5% containing *n*-butanol. It may be noted that the largest reduction in PM emissions was observed for fuels with the highest oxygen content. Among fuels with a similar oxygen content, the most effective in emission reduction were components with lower cetane number. It can also be pointed out that physical properties of the PT-1-5% component, deviating from those of the diesel fuel, were not an obstacle in PM emission reduction.

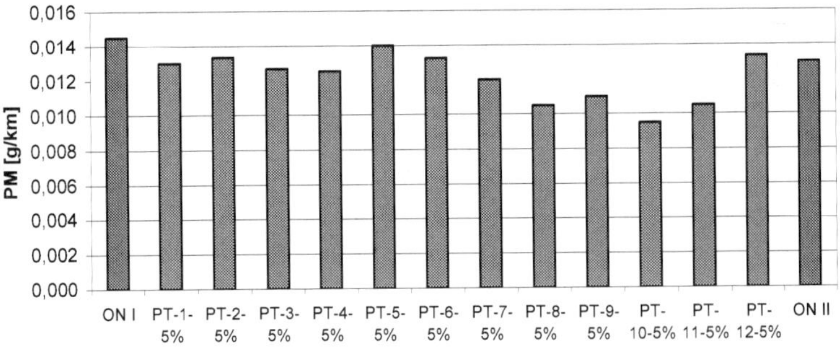

Figure 16. PM emissions during the UDC for conventional diesel fuels and oxygenated fuels containing individual oxygenates at a concentration of 5% v/v

During the EUDC phase, the impact of oxygenated fuels on emissions of PM was even more favorable and clear (Fig. 17) than in UDC. Classification of oxygenates in terms of efficacy was similar to the one in the UDC phase, where the most effective component (11) lowered PM emissions by 40%.

PM emission level when running on diesel fuel was higher in the EUDC phase than in the UDC phase, whereas for oxygenated fuels, it was roughly similar in both phases, and therefore the

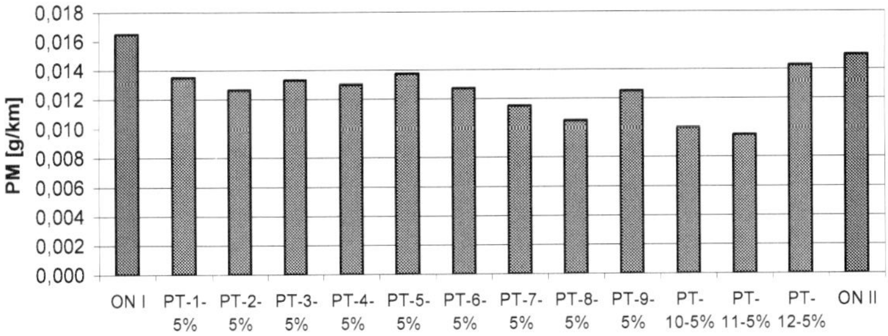

Figure 17. PM emissions during the EUDC for conventional diesel fuels and oxygenated fuels containing individual oxygenates at a concentration of 5% v/v

application of oxygenated fuels resulted in a greater reduction in PM emissions in the EUDC phase. On average, over the whole NEDC, oxygenated fuels achieved at least a few percent of reduction in PM emissions (Fig. 18).

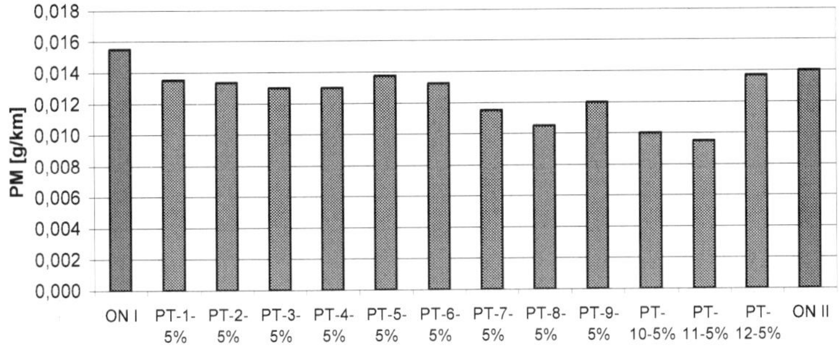

Figure 18. The average PM emissions in the NEDC (UDC + EUDC) for conventional diesel fuels and oxygenated fuels containing individual oxygenates at a concentration of 5% v/v

Glycol ethers, which for the previously analyzed exhaust gas components cause the most favorable changes, were less effective for PM emissions. The greatest reduction in PM emissions was obtained for carbonates and for maleates. The use of *n*-butanol resulted in the lowest reduction in PM emissions of the tested fuels with oxygenated compounds.

3.3.5. CO_2 emissions and fuel consumption

CO_2 emissions (Fig. 19) and fuel consumption (Fig. 20) in the NEDC was only slightly higher – in the range of measurement uncertainty – for oxygenated fuel.

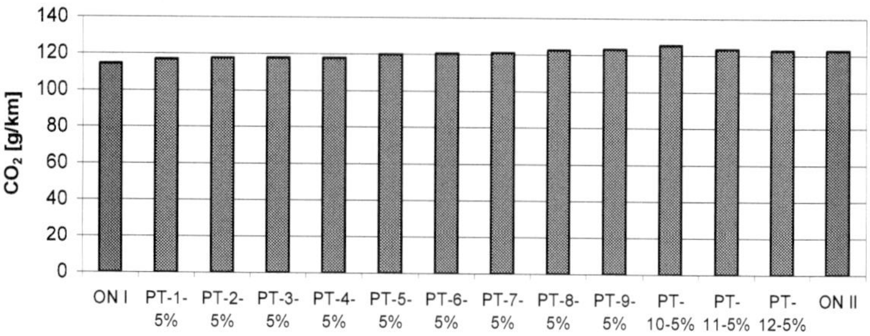

Figure 19. The average CO_2 emissions in the NEDC (UDC + EUDC) for conventional diesel fuels and oxygenated fuels containing individual oxygenates at a concentration of 5% v/v

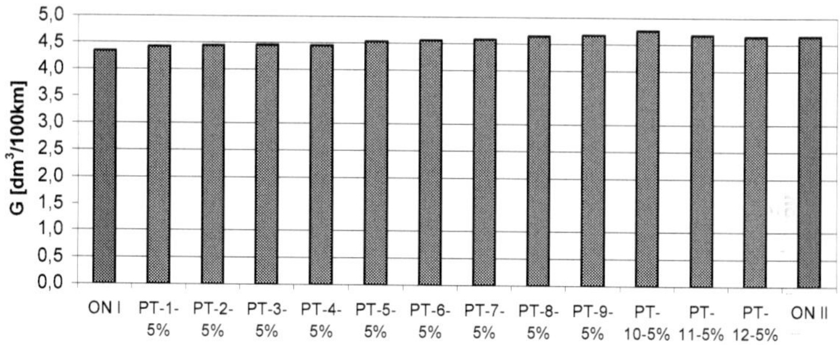

Figure 20. The average fuel consumption in the NEDC (UDC + EUDC) for conventional diesel fuels and oxygenated fuels containing individual oxygenates at a concentration of 5% v/v

3.3.6. A summary of preliminary results

Figures 21–24 show the percentage changes in exhaust emissions of the test vehicle in the NEDC by using fuels containing individual oxygenates at a concentration of 5% v/v. The results indicate that some of the oxygenated fuels caused a reduction of CO and HC, and all the fuel types tested caused a reduction of emissions of PM and an increase in NO_x emissions.

It should be emphasized that the scale of emission changes of individual exhaust components caused by the different oxygenated fuels varied. It was generally found that fuels containing oxygenated compounds from the group of glycol ethers resulted in the largest, of up to 40%, reduction in emissions of CO and HC with an increase of several percent in NO_x emission and a moderate reduction of PM emission. Fuels containing carbonates provided the greatest, even by more than 30%, reduction in PM emissions; however, they also increased the emissions of other toxic components of exhaust gases. The advantages of fuels containing maleates were no increase in NO_x and PM reduction greater than when using glycol ethers, although using maleates increased emissions of CO and HC. The effect of oxygenated fuel containing *n*-

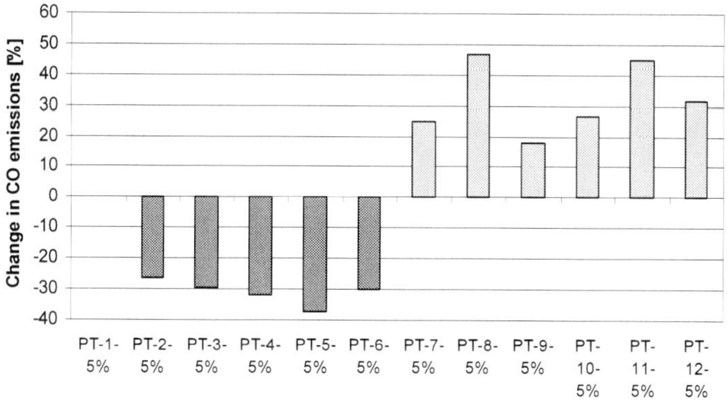

Figure 21. The percentage change in CO emissions in the NEDC (UDC + EUDC) for oxygenated fuels containing individual oxygenates at a concentration of 5% v/v with respect to regular diesel fuel

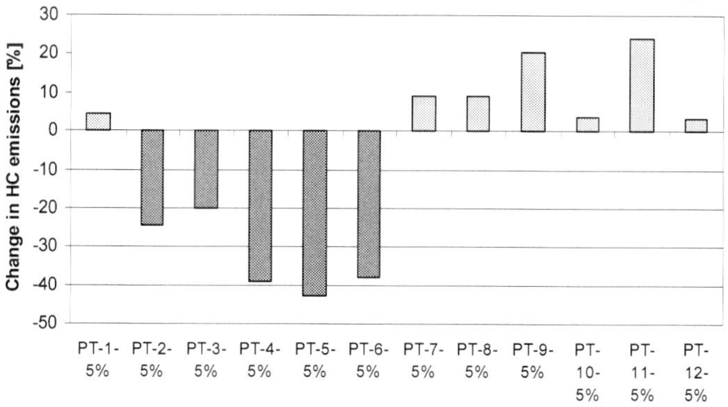

Figure 22. The percentage change in HC emissions in the NEDC (UDC + EUDC) for oxygenated fuels containing individual oxygenates at a concentration of 5% v/v with respect to regular diesel fuel

butanol on the exhaust emissions is rated as unfavorable, because it resulted in a significant increase in CO emissions at a much lower decrease in PM emissions than for other fuels used.

The literature on the influence of oxygenates on the PM emissions for CI engines contains diverging opinions on whether the reduction in PM emissions depends only on the amount of oxygen introduced to the fuel [21, 26, 32, 33], or also on the type of the oxygenated compound, which is to be used [17, 23, 27, 29]. This divergence of opinion is based on the fact that very few papers [14, 24] present the results of using a wide variety of oxygenates under identical measurement conditions. Comparison of the change of PM emissions in the NEDC shown in Fig. 25 to the oxygen content in the individual oxygenated fuels indicates that 1% of the oxygen contained in the fuel corresponds in most of the cases to a reduction in emissions

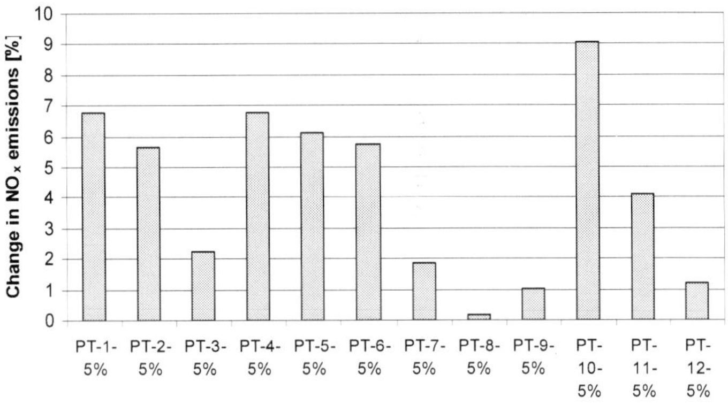

Figure 23. The percentage change in NO_x emissions in the NEDC (UDC + EUDC) for oxygenated fuels containing individual oxygenates at a concentration of 5% v/v with respect to regular diesel fuel

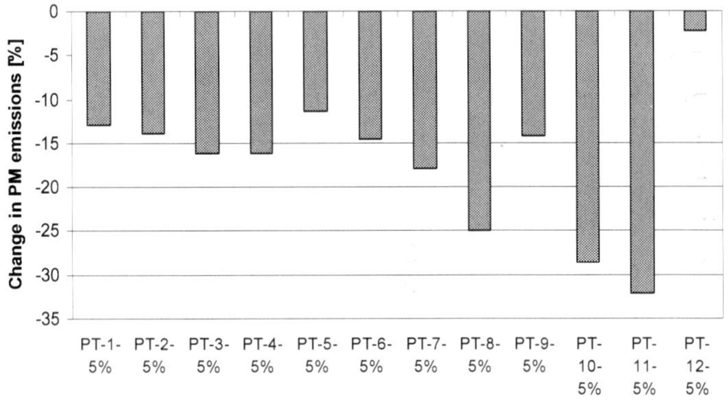

Figure 24. The percentage change in PM emissions in the NEDC (UDC + EUDC) for oxygenated fuels containing individual oxygenates at a concentration of 5% v/v with respect to regular diesel fuel

of PM in the NEDC by 7–10%. It can be concluded that the oxygen content of fuel is a very important, but not the only, parameter determining the level of reduction of PM emissions.

In assessing the impact of individual oxygenates on the exhaust emission, what should be taken into account, first of all, are the changes in the emission of PM and NO_x, including the mutual relation of these changes (i.e. PM/NO_x trade-off), due to the fact that they are, on the one hand, the most harmful components, and on the other, the most difficult to remove from the exhaust gas. The effect on emissions of CO and HC should only be taken into account in the longer term, particularly if the impact is significant. On this basis, six oxygenates (shown in the Table 6), which resulted in the most favorable changes in exhaust emissions during preliminary tests described in this chapter, were selected for further, more detailed studies.

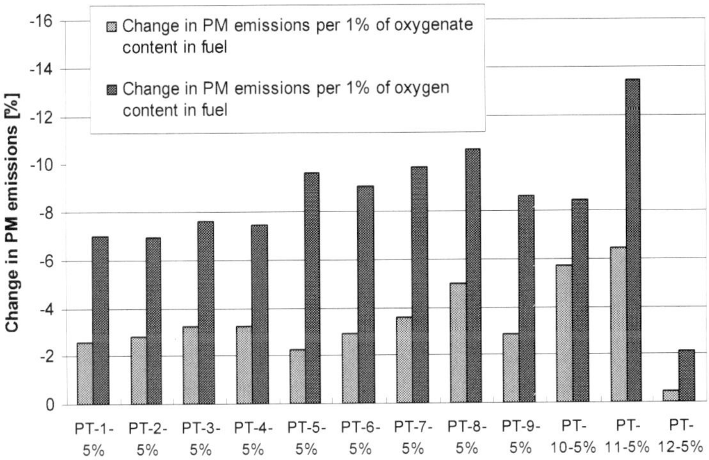

Figure 25. The relative effect of oxygen compounds and oxygen content of the fuel on PM emissions in the NEDC

Oxygenate no.	Chemical name
3	Triethylene glycol dimethyl ether
4	Tetraethylene glycol dimethyl ether
5	Diethylene glycol dibutyl ether
6	Dipropylene glycol dimethyl ether
8	Diethyl maleate
11	Diethyl carbonate

Table 6. Oxygen compounds selected for further, more detailed analysis

3.4. Evaluation of the impact of oxygenated compounds based on their content in the fuel

The relationship between the changes in emissions and the content of oxygenated compounds in the fuel is a key issue from the point of view of rational use of oxygenates. Recognition of these relationships also provides additional data for the analysis of the mechanisms of the effect of oxygenates on emissions of toxic exhaust components. The literature reports on the relationship between PM emissions and the oxygen content in the fuel, which is usually defined as approximately linear [20, 26]. Figure 26 shows the results of the influence of the six selected oxygenates used at concentrations of 5 and 10% v/v on the change in emissions of CO, HC, NO_x, and PM in the NEDC.

Analyzing these results reveals different impacts of fuels containing components from different chemical groups. In particular, the different impact on emissions between fuels containing glycol ethers (components 3–6), maleates, or carbonates (components 8 and 11).

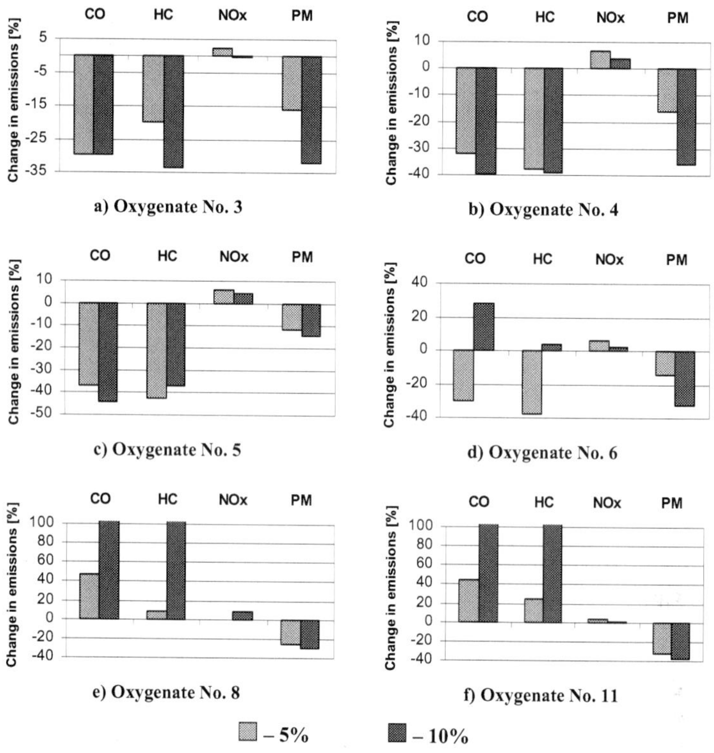

Figure 26. The impact of oxygenate content on the change in exhaust emissions over the NEDC

In the case of glycol ethers, it can generally be said that the amount of oxygenated compounds, and thus the amount of oxygen in the fuel, affects PM emissions the most. The reduction of this emission is generally proportional to the content of the oxygenated component/oxygen. Doubling the content of components 3, 4, and 6 resulted in twice the reduction in PM emissions. Increasing the content of the oxygenated components also resulted in the reduction of NO_x emission compared to fuels containing a smaller amount of the oxygenate. Taking into account the PM/NO_x relationship, an increase in the levels of these oxygenated components from 5 to 10% produced clearly beneficial effects.

For fuels containing components 8 and 11, doubling the oxygenated component content in the fuel did not cause a significant reduction in PM emissions. At 5% content, the glycol ethers were clearly less effective in reducing PM emission than diethyl maleate and diethyl carbonate. At 10% content, the efficiency of the best components from all tested chemical groups was similar – PM reduction of 30–40%. For most of these fuels, favorable NO_x emission values at a higher content of oxygenated component were also reported.

In relation to CO and HC, it can be generally stated that the oxygenated compounds from the group of glycol ethers were effective in reducing CO and HC emissions in NEDC, while increasing the concentration of the individual oxygenates caused various effects. For component 3, a significant reduction in HC emissions at the same level of CO emission (Fig. 26a) was achieved. For component 4, CO emission was decreased by an additional 8 percentage points, with the same level of HC emission (Fig. 26b). For component 5, CO emission was reduced by a few extra percentage points (Fig. 26c), while increasing emissions of HC in a similar range. For component 6, an increase in CO and HC emissions was reported (Fig. 26d) at higher concentrations of oxygenated components.

Emissions of CO and HC for fuels containing components 8 and 11 (Fig. 26e and f) were clearly higher than for neat diesel, and increasing the oxygen content of the fuel greatly increased this unfavorable phenomenon. For a content of 10% of the oxygenated components 8 and 11, CO and HC emissions were over 100% higher than for neat diesel.

In the light of these results, it can be concluded that the use of oxygenated components in the concentration of 10% is preferred especially for components 3 and 4, where the higher content in the fuel meant significantly lower emissions of PM, and the same or more favorable emission levels of other toxic exhaust components as for the 5% content.

3.5. The effect of oxygenates on exhaust emissions in the conditions of different test cycles

The level of exhaust emissions for a vehicle of given emission class largely depends on the operating conditions of its engine, in particular parameters such as speed, load, thermal state, as well as the dynamics of changes in these operating conditions. To assess the impact of engine operating conditions on the reduction in exhaust emissions as a result of addition of oxygenated compounds, the previously performed emission tests in the NEDC regime were followed up with the same type of research carried out for the FTP-75 cycle. Despite being the most widely used worldwide cycles in research of exhaust emissions, the conditions of performing these two cycles are significantly different. Both drive cycles are described in more detail in Section 3.2, but at this point, it should be emphasized that the FTP-75 cycle is more dynamic than NEDC. It is characterized by a more aggressive acceleration, smaller share of engine idling and maximum speed for the cycle (which is slightly lower than for the NEDC) reached in its early stages already.

Figure 27 shows, as an example of the tests performed, a comparison of the results of measurements of exhaust emissions in the individual phases and the average over the FTP-75 cycle for the test vehicle powered by diesel fuel and oxygenated fuel PT-3-10%. It can be seen that the high CO emission occurs only in the first phase of the FTP-75 cycle. In further phases, the engine thermal condition is stabilized, and the emission significantly reduced. HC emissions are also highest in the first phase, but the decrease in emissions in subsequent phases is not as significant as for CO. This can be related to different efficiencies of catalytic converter in relation to CO and HC emission reduction. It should be pointed out that the exhaust gas temperature in the different phases of the FTP-75 cycle was lower than in the EUDC phase of the NEDC. NO_x emissions during all phases reached similar values, while for PM emissions,

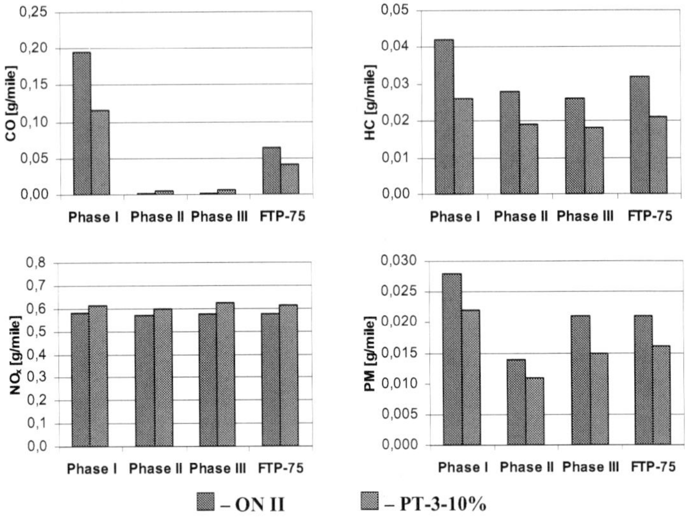

Figure 27. Exhaust emissions in the various phases and the average over the FTP-75 cycle for the test vehicle powered by diesel fuel and oxygenated fuel PT-3-10%

there is a clear dependence on the thermal state and the engine load in the different phases of the FTP-75 cycle.

Figure 28a–f compares the change in exhaust emissions over the NEDC and FTP-75 test cycles as a result of application of oxygenated fuels: PT-3-10%, PT-4-10%, PT-5-10%, PT-6-10-%, PT-8-10%, and PT-11-10%, respectively. Analysis of the results indicates that when there was a significant change in emissions (over 10%), then the direction of emission changes caused by the use of the oxygenated fuel was the same for both test cycles. For most oxygenated fuels, a similar scale of changes in CO and HC emissions was achieved in both test cycles. Changes in NO_x emissions for all oxygenated fuels in both cycles were small – less than 10%. Thus, only PM emissions can point to a clear link between the level of emission changes caused by the use of oxygenated fuels and the conditions of the test cycle. For all the examined oxygenated fuels, and for most significantly, a greater reduction in PM emissions was achieved in the NEDC than the FTP-75 cycle conditions. Larger maximum load on the engine found in the EUDC phase is thought to have been a significant factor.

4. Conclusions

1. The use of oxygenated fuels has a significant impact on the level of exhaust emissions from diesel engines without particulate filters. This effect is significant even at low oxygenate content in the fuel (5% v/v).

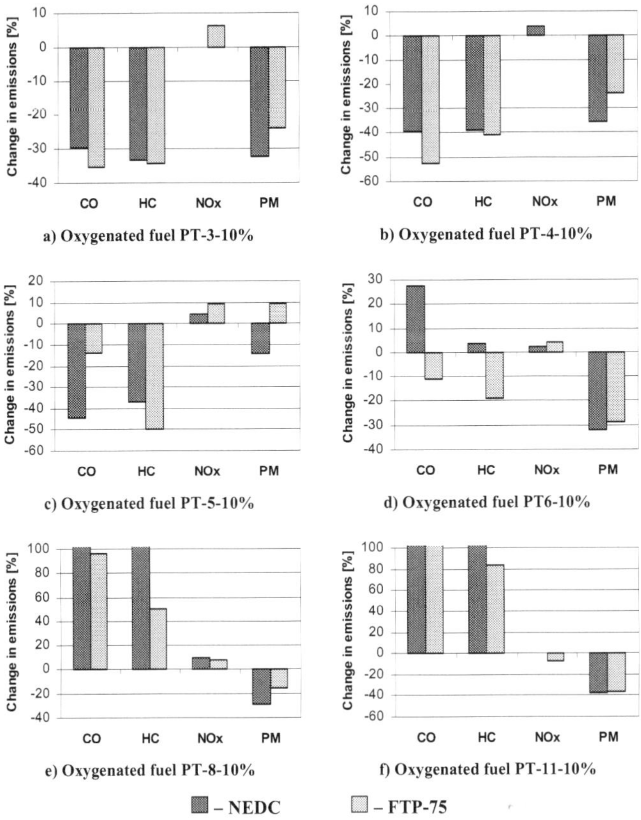

Figure 28. Comparison of changes in exhaust emissions as a result of application of oxygenated fuels over the NEDC and the FTP-75 cycle conditions

2. When using oxygenated fuels, irrespective of the component, a significant reduction in PM emissions and a slight increase in NO_x emissions were found in the both NEDC and FTP-75 cycle. Changes in CO and HC emissions depend on the type of oxygenated compound used.

3. The main parameter that determines the degree of reduction in PM emissions is the oxygen content in the fuel. After using various oxygenates at a concentration of 5% v/v, it was found that 1% of the oxygen present in the fuel caused the PM emissions of a Euro 4 class vehicle in NEDC to be reduced by 7–10%.

4. The type of an effect that oxygenated fuels have on CO and HC emissions is related to oxygen content to a lesser extent, and mainly depends on their autoignition properties.

Fuels containing glycol ethers cause a reduction of CO and HC emission, while fuels containing maleates, carbonates or butanol cause its growth.

5. Increasing the concentration of oxygenates in the fuel from 5 to 10% v/v causes different effects for different components. The most favorable changes in emissions were observed for glycol ethers – where increased concentration was usually followed by a further significant reduction in PM emissions, and, to a lesser extent, reductions in CO and HC emissions. For maleates and carbonates, changes were less favorable, because a slight decrease in PM emissions was accompanied by a significant increase in CO and HC emissions.

6. The effect of the test conditions (NEDC/ FTP-75 cycle) on changes in exhaust emissions occur primarily for PM – where a greater reduction was achieved in the NEDC. Changes in NO_x emissions for all fuels in both cycles do not exceed 10%.

7. The slight increase in NO_x emissions and substantially lower PM emissions associated with the use of oxygenated fuels should be judged as a favorable change in the correlation of PM/NO_x emissions (the PM/NO_x trade-off). In this situation, there is a possibility for reducing the exhaust emissions of both aforementioned components at the same time.

Author details

Miłosław Kozak*

Address all correspondence to: Miloslaw.Kozak@put.poznan.pl

Poznań University of Technology, Poznań, Poland

References

[1] European Environment Agency, www.eea.europa.eu, access on 15 June 2015.

[2] United States Environmental Protection Agency (EPA): Health Assessment Document For Diesel Engine Exhaust, Waszyngton, 2002.

[3] European Automobile Manufacturers' Association (ACEA), www.acea.be, access on 15 June 2015.

[4] Główny Urząd Statystyczny: Gospodarka paliwowo-energetyczna w latach 2011, 2012, Warszawa 2013 (*Polish Central Statistic Office: Energy Statistics 2011, 2012, Warsaw 2013*).

[5] Benazzi E.: Gasoline and Diesel Imbalances in the Atlantic Basin. Petroleum Technology Quarterly, nr 2/2011.

[6] Merkisz J., Kozak M.: Internal Combustion Engines at the Beginning of the Third Millennium. International Scientific Conference: Development of Design of Internal Combustion Engines and Quality of Fuels to Meet Future Levels, Kraków 20-21.06.2011.

[7] Rutherford J.A.: Effects of Gasoline Properties on Emissions of Current and Future Vehicles – T50, T90, and Sulfur – Auto/Oil Air Quality Improvement Research Program. SAE Paper 952510, 1995.

[8] Tsuda H., Ito T., Nakamura K.: Japan Clear Air Program (JCAP) – Program Objectives and Design. SAE Paper 1999-01-1481.

[9] MacKinven R., Hublin M.: European Program on Emissions, Fuels and Engine Technologies – Objectives and design. SAE Paper 961065, 1996.

[10] Rickeard D.J., Bonetto R., Singer M.: European Programme on Emissions, Fuels and Engine Technology (EPEFE) – Comparison of Light and Heavy Duty Diesel Studies. SAE Paper 961075, 1996.

[11] Murphy M.: Oxygenate Compatibility with Diesel Fuels. SAE Paper 2002-01-2848.

[12] Natarajan M. et al.: Oxygenates for Advanced Petroleum-Based Diesel Fuels: Part 1. Screening and Selection Methodology for the Oxygenates. SAE Paper 2001-01-3631.

[13] Bertola A., Boulouchos K.: Oxygenated Fuels for Particulate Emissions Reduction in Heavy-Duty DI-Diesel Engines with Common Rail Fuel Injection. SAE Paper 2000-01-2885.

[14] Delfort B., Durand I., Jaecker-Voirol A., Lacome T., Paille F., Montagne X.: Oxygenated Compounds and Diesel Engine Pollutant Emissions Performances of New Generation of Products. SAE Paper 2002-01-2852.

[15] Murphy M., Lanning L.: Vapor Pressures of Diesel Fuel Oxygenate Blends. SAE Paper 2002-01-2850.

[16] Murphy M.: Safety and Industrial Hygiene Issues Related to the Use of Oxygenates in Diesel Fuels. SAE Paper 1999-01-1473.

[17] Stoner M., Litzinger T.: Effects of Structure and Boiling Point of Oxygenated Blending Compounds in Reducing Diesel Emissions. SAE Paper 1999-01-1475.

[18] Gonzalez M.A., Piel W., Asmus T., Clark W., Garbak J., Liney E., Natarajan M., Neageli D.W., Yost D., Frame E.A., Wallace J.P.: Oxygenates Screening for Advanced Petroleum-Based Diesel Fuels: Part 2. The Effect of Oxygenate Blending Compounds on Exhaust Emissions. SAE Paper 2001-01-3631.

[19] Gonzalez M.A., Clark W., Wolf L.R., Garbak J.A., Wright K.J., Natarajan M., Yost D.M., Frame E.A., Kenney T.E., Ball J.C., Wallace J.P., Hilden D.L., King D.: Impact of Engine Operating Conditions on Low-NO_x Emissions in a Light-Duty CIDI Engine Using Advanced Fuels. SAE Paper 2002-01-2884.

[20] Murayama T. et al.: Simultaneous Reductions of Smoke and NO_x from a DI Diesel Engine with EGR and Dimethyl Carbonate. SAE Paper 952518, 1995.

[21] Akasaka Y., Sakurai Y.: Effects of Oxygenated Fuel and Cetane Improver on Exhaust Emission from Heavy-Duty DI Diesel Engines. SAE Paper 942023, 1994.

[22] Kocis D., Song H., Lee H., Litzinger T.: Effects of Dimethoxymethane and Dimethyl-carbonate on Soot Production in an Optically-accessible DI Diesel Engine. SAE Paper 2000-01-2795.

[23] Hallgren B., Heywood J.: Effects of Oxygenated Fuels on DI Diesel Combustion and Emissions. SAE Paper 2001-01-0648.

[24] Yeh L., Rickeard D., Duff J., Bateman J., Schlosberg R., Caers R.: Oxygenates: An Evaluation of their Effects on Diesel Emissions. SAE Paper 2001-01-2019.

[25] Porai P., Chandrasekaran S., Subramaniyam S., Jancirani J., Sahoo B.: Combustion and Performance of a Diesel Engine with Oxygenated Diesel Blend. SAE Paper 2004-01-0082.

[26] Nabi M., Minami M., Ogawa H., Miyamoto N.: Ultra Low Emission and High Performance Diesel Combustion with Highly Oxygenated Fuel. SAE Paper 2000-01-0231.

[27] Mueller C.J., Martin G.C.: Effects of Oxygenated Compounds on Combustion and Soot Evolution in a DI Diesel Engine: Broadband Natural Luminosity Imaging. SAE Paper 2002-01-1631.

[28] Buccholz B.A., Mueller C.J., Upatnieks A., Martin G.C., Pitz W.J., Westbrook, C.K.: Using Carbon-14 Isotope Tracing to Investigate Molecular Structure Effects of the Oxygenate Dibutyl Maleate on Soot Emissions from a DI Diesel Engine. SAE Paper 2004-01-1849.

[29] Boot M., Frijters P., Klein-Douwel R., Baert R.: Oxygenated Fuel Composition Impact on Heavy-Duty Diesel Engine Emissions. SAE Paper 2007-01-2018.

[30] Bielaczyc P., Kozak M., Merkisz J.: Effects of Fuel Properties on Exhaust Emissions From the Latest Light-Duty DI Diesel Engine. Advanced Emission Aftertreatment Systems. SAE International, SP-1796, 2003.

[31] Kozak M.: Wpływ wybranych parametrów paliwa na emisję toksycznych składników spalin z silników o zapłonie samoczynnym. Praca doktorska, Politechnika Poznańska, Poznań 2004. (*Kozak M.: The Influence of Some Diesel Fuel Properties on Exhaust Emissions from Diesel Engines. PhD Thesis, Poznan University of Technology, Poznan 2004*).

[32] Miyamoto N., Ogawa H., Nurun N., Obata K., Arima T.: Smokeless, Low NO_x, High Thermal Efficiency, and Low Noise Diesel Combustion with Oxygenated Agents as Main Fuel. SAE Paper 980506, 1998.

[33] Tree D., Svensson K.: Soot Processes in Compression Ignition Engines. Progress in Energy and Combustion Science vol. 33 (2007).

Chapter 4

Combustion of Gaseous Alternative Fuels in Compression Ignition Engines

Mikulski Maciej, Wierzbicki Sławomir, Ambrosewicz-Walacik Marta, Duda Kamil and Piętak Andrzej

Additional information is available at the end of the chapter

http://dx.doi.org/10.5772/61663

Abstract

The problem of alternative fuels for combustion engines has been growing in importance recently. This is connected not only with decreasing fossil fuel resources, but also with the growing concern for the natural environment and the fight against global warming. This paper discusses the possibility of utilizing alternative gaseous fuels in compression-ignition engines, using dual-fuel, gas-liquid operation strategy. Current state of the art of this technology had been introduced, along with its benefits and challenges to be countered. The discussion had been supported by authors own research experience on dual-fuel engines. The latest results of research on the impact of gas composition on combustion process in the Common Rail dual fuel engine had been presented, at the same illustrating the environmental benefits of using gaseous fuels. The Utilization of gaseous fuels with varying composition was illustrated systematically, starting with natural gas. The possibility of using fuels with lower content of methane (the so-called low-calorie gases) was shown by the impact of depleting natural gas with carbon dioxide. Industrial gases, such as syngas contain a large amount of hydrogen, carbon monoxide or higher hydrocarbons (ethane, propane). The possibility of fueling CI engines with these gasses was presented by the influence of enriching natural gas with mentioned components. The results cover engine dynamometer tests for different operating conditions with the analysis of the combustion process and detailed emission measurements discussion. The results of experimental studies were supplemented by simulation results, using mathematical models, developed by the authors for multi-fuel enginesr.

Keywords: Dual-fuel, Compression Ignition, Common-Rail, Gaseous Fuels, Natural Gas, Emissions, Performance

1. Introduction

Currently, over 20% of global demand for energy is managed by transport [3], which to a significant extent, has been dominated by internal combustion engines. Heavy duty road transport, maritime transport and almost half of passenger cars are powered by compression-ignition (CI) engines. Increasingly, internal combustion engines have been used for energy generation.

Higher efficiency compared to spark ignition engines and the ability to operate on a wide range of fuels predisposes the modern CI engine to be considered as one of the powertrains of the future. It should be noted that modern combustion technologies available with the dissemination of electronic fuel injection process control and a range of new concepts for exhaust aftertreatment control have made CI engines become more and more environmentally friendly. Simultaneously, wide spectrum of regulations results in clear and efficient conversion of chemical energy contained in the fuel, provided for different carriers of this energy. This flexibility allow modern CI engine to operate on liquid fuels with extremely different physicochemical parameters without the need for structural modifications. The optimization of engine operation can be done at the level of control algorithms [2].

Recent years there has been a significant increase in interest in gaseous fuels. Their share in the global energy balance will soon exceed 21% of the total energy produced [4]. This is due to the fact that regardless of the type of gas and the applied combustion technology, these fuels allow for a significant reduction in toxic emissions. On the other hand, very high availability (deposits of natural gas, shale gas, ability to obtain bio-methane from organic matter, syngas and coal gas from industrial processes) and low price cause that, in the coming years, gaseous fuels will significantly diversify all energy sectors.

Currently, many research centers are developing the concept of powering CI engines with gaseous fuels in a dual-fuel mode. A relatively small dose of liquid fuel, injected into the cylinder, acts as an ignition inhibitor and stabilizes the combustion of gaseous fuel which provides major part of energy. As in the case of mono-operation, the gaseous fuel supplied to the engine can change. Adaptation to the changes in fuel composition is reached by regulating the fuel injection parameters of initial dosage and many other regulation parameters like air amount and temperature, exhaust recirculation, etc.

Appropriately designed CI engine with complex regulation algorithms (often based on artificial neural networks and advanced mathematical models) can essentially be called a multi-fuel engine which can be powered at any time with any liquid or gaseous fuel depending on availability.

For more than 7 years, our team has been carrying out projects in the field of utilization of gaseous fuels in compression ignition engines. In the following chapter the results of dose works had been summarized introducing the reader to the topic of multi-fuel, gas –liquid, engines.

1.1. Gaseous fuels for CI engines

Limited supplies of fossil fuels and the fight against global warming and greenhouse gas emissions all contribute to the search for new types of fuels that could be effectively used in modern compression-ignition engines [7, 38, 40, 43, 50].

Some properties of the flammable gases that are the basic constituents of gaseous fuels are shown in table 1 [7, 38, 40].

Component name	Calorific value [MJ/m³ n]	Density at normal conditions [kg/m³]	Lower ignition limit (% of gas in air)	Upper ignition limit (% of gas in air)	Speed of combustion [m/s]	Auto-ignition temperature [K]	Theoretical air requisition [m³/m³]
Hydrogen H_2	10,78	0,0899	4	75	0,302	807	2,38
Methane CH_4	35,89	0,717	5	15	0,338-0,67	923	9,54
Ethane C_2H_6	63,77	1,34	3	12,4	0,43-0,856	793	16,7
Propane C_3H_8	91,28	1,97	2,1	9,5	0,384-0,821	783	23,8
Carbon monoxide CO	12,6	1,25	12,5	74	0,024	881	2,38
Hydrogen sulfide H_2S		1,54	4,3	45	-	563	7,14

Table 1. The basic properties of the individual components of gaseous fuels [7, 38, 40].

Of all the above-mentioned flammable gases only methane is a fossil gas. The remaining gases are obtained as a result of the processing of minerals, or via autogenous, natural processes.

Methane is the main component of natural gas. Its content varies depending on the gas source, and is usually between 90 and 99%. Hard coal mines are another source of methane, where significant amounts of this gas are obtained via the so-called de-methanising process of the mines. The main advantage of using methane as combustion engine fuel is its low coal content, which means that its combustion results in much lower carbon dioxide emission levels. Moreover, it is characterised by a very high octane number (125-130), which allows it to be used in engines with a higher compression ratio, and ensures improved general engine efficiency [40, 50]. Natural gas is usually stored in compressed form as CNG (Compressed Natural Gas). It is also possible to store this gas in liquefied form as LNG (Liquefied Natural Gas), but this requires very low storage temperatures of the order of -163 deg. C [43].

Of the remaining gas fuels available, of significance currently are liquefied petroleum-derivative LPG (Liquefied Petroleum Gas) gases, which are a mixture of propane and butane.

Currently, however, particular attention is given to gas fuels obtained from renewable energy sources. Gas fuels obtained from various sorts of biomass, animal excrement, communal waste, sewage, and agricultural/food industry waste are defined as biogas [7, 43, 50].

The composition of biogas is not permanent, and depends not only on the raw materials that it is made of, but also on the technology used for its production. The estimated chemical composition of biogas made according to different methods and materials is shown in table 2 [7].

Component	Content		
	Agricultural biogas	**Treatment plant biogas**	**Landfill biogas**
Methane CH_4	45-75 %	57-62 %	37-67 %
Carbon Dioxide CO_2	25-55 %	33-38 %	24-40 %
Oxygen O_2	0,01-2,0-2,1 %	0-0,5 %	1-5%
Nitrogen N_2	0,01-5,0 %	3,4-8,1 %	10-25 %
Hydrogen sulfide H_2S	10-30 000 ppm	24-8 000 ppm	15-427 ppm

Table 2. Composition of biogas, depending on its origin [7].

During the analysis of the properties of the aforementioned flammable gases in view of their use in self-ignition engines, above all attention must be paid to their relatively high self-ignition temperatures. Another downside of the use of these fuels in self-ignition engines is the relatively narrow limit of their flammability, which in combination with the relatively high global excess air coefficient for these engines restricts gas combustion in the combustion chamber [40, 43].

Unprocessed, directly drawn biogas is heavily contaminated, and contains significant amounts of non-flammable components. Moreover, the hydrogen sulfide that naturally resides in biogas has highly corrosive properties. The methods of treatment of biogas have been well recognised [3, 21], and the need for its purification does not represent a significant restraint to its use on a mass scale. As a result of the treatment process, proper engine fuel, called biomethane, is obtained. After treatment, biogas has a similar composition to natural gas, and therefore has very similar physicochemical properties.

Hydrogen has a very small molecular weight and a high gross calorific value. It has the highest energy-to-mass ratio among the known hydrocarbon fuels, maintaining at the same time beneficial physicochemical parameters in view of fuel combustion in gas engines. The main advantages of this fuel include the lack of emission of carbon dioxide during combustion. The use of hydrogen as an alternative gas fuel is currently restricted by the low efficiency of the process of its extraction. Currently, the production of hydrogen via electrolysis is 3.5 times more expensive than its extraction using catalytic steam reforming of fossil fuels [44]. In molecular form, hydrogen can also be manufactured using the natural gas reforming process. The direct combustion of natural gas is still more effective than its conversion into hydrogen fuel. An extensive analysis of the methods of production of hydrogen has been shown in [13, 30, 44]. The use of hydrogen as fuel for engines is also limited by the high reactivity of the gas, which restricts its storage and transport capabilities, and enforces the use of special non-reactive alloys. Additional constraints are related to the low density of the gas, which imposes special requirements regarding the sealing in engine components. Due to the above, clean molecular hydrogen is not a realistic alternative to petroleum-derivative fuels as engine fuel for common use. It must be stressed that there are new concepts for the production of molecular hydrogen, and the highest expectations are currently invested in the use of biological processes - anaerobic fermentation [5, 46] and photolysis of water [31, 37], which are still not fully developed as of today. However, disregarding the development of the technology for the production of hydrogen in the future, it is estimated [4] that the main use of this fuel will be limited to the storage of energy in the form of fuel cells, and as additives to other gas fuels.

Currently, a new technology for gas fuels, HCNG, is being developed. The abbreviation stands for natural gas (or biomethane), enriched with hydrogen to the level of 4-9% of the total volume. Research on these kinds of compositions is being carried out mainly in the United States. The main advantage of HCNG is the ability to significantly reduce the toxicity of emissions in relation to pure CNG. Karner and Frnackfort tests [17] for using this fuel to power spark ignition engines adapted to the use of CNG have shown that compounds below 20% do not require additional adapting or changing of the control programme. It was also proved that a 50% hydrogen content level does not increase the probability of explosion as compared to pure CNG, and can be used with appropriately regulated engine parameters. Adding hydrogen to fuel increases its calorific value. During the aforementioned tests, the engine achieved 5% more power than when powered by pure CNG.

1.2. Advanced combustion and control technologies for alternative-fuel CI engines

Along with the search for new, cleaner fuels for combustion engines there is a need to increase the efficiency of conversion of chemical energy contained in the fuel, with particular emphasis on real engine's emissions.

A milestone for improving the combustion of liquid fuels in CI engines was the introduction of electronic injection process control. Common Rail Direct Injection (CRDI) gradually replaces traditional, mechanically controlled injection systems in all CI Engine developments, from light-duty to heavy-duty stationary and marine engines. In the electronic Common Rail (CR) injection systems that are commonly used nowadays, it is possible to precisely adjust the injection timing, pressure and duration. Furthermore, CR systems also have the option of dividing the fuel into several doses, injected sequentially, depending on the engine operation conditions. In such systems, usually a very small portion of diesel (pilot dose) is injected early in the compression phase to improve the conditions for the main dose combustion, which is injected some time after the ignition of the pilot dose. Divided injection enables additional control over the combustion process, reduces toxic emissions and noise levels [29].

New control possibilities provided by the introduction of CR injection systems do not cover all potential of CI engines, related to alternative fuels utilization. New modes of combustion are being sought in order to reduce emission levels and fossil fuel consumption. A potential candidate is the Homogeneous Charge Compression Ignition (HCCI) engine concept. In the HCCI engine a premixed air/fuel mixture is introduced to the cylinder. Combustion takes place spontaneously when the homogenous fuel mixture has reached its chemical activation energy and therefore is controlled by chemical reaction kinetics only [18].

The HCCI technology isn't mature enough to be commercially available yet, but it is being intensively developed [54]. Diesel fuel alone is no suitable for HCCI operation due to its low volatility and high propensity. Gasoline HCCI engine concepts have been studied by many researchers [10, 33, 49, 57], showing the possibility to cut fuel consumption by a maximum of 15% while virtually eliminating NOx emissions. Still, stable high load operation is the main issue to be solved for gasoline HCCI engines. One of the possible solutions is to blend gasoline with fuels with high knock resistance properties (diesel or liquid biofuels) [6].

The combination of different gaseous fuels (natural gas, hydrogen, propane) with diesel in HCCI mode is reported to yield low emissions and, to some extent, increase engine efficiency [36, 47]. The same time it is possible to increase alternative fuel energy share even further by substituting diesel fuel with liquid biofuels. The combination of both high cetane number liquid fuel and high octane – gaseous fuel can provide a soft engine run while operating at high compression ratios [11]. It was found that high compression ratio (up to 18:1) engine has an advantage of producing ultralow NOx emissions and high thermal efficiency at steady state operation [15].

The problems with proper control of the combustion process, had caused that the HCCI gas-diesel operation, have consequently evolved to another dual-fuel concept called PCCI (Premixed Charge Compression Ignition) or RCCI (Reactivity Controlled Compression Ignition) [8, 12, 22]. The concept combines the advantages of HCCI combustion while introducing CR direct injection of liquid fuel. It is assumed that small amount of diesel fuel is injected early during the compression phase and therefore can be completely premixed before the auto ignition starts. This keeps the advantages of HCCI operation while, by in cylinder blending, showing good potential in terms of controllability and high load operation. The research on RCCI has focused mostly on the combination of gasoline-like and diesel-like fuels. For the combination of natural gas and diesel limited studies are available. Especially full scale engine test results with such operation are lacking.

It is worth to notice that this technology has a good potential for using a wide spectrum of fuels with significantly different composition and thermal/energy properties. This leads to the concept of a multi-fuel gaseous engine. In multi-fuel engines, currently recognised fuelling methods should be subject to modification. The share of individual fuel types in total energy content should be defined in order to ensure maximum reduction of toxicity of emissions and achievement of maximum engine efficiency. To ensure optimum utilisation of the engine's capabilities, these parameters should be regulated depending on engine speed and load. With such a defined fuelling strategy the selection of the right control algorithm is not a simple task. Modern self-ignition engines with electronic control systems ensure simpler regulation. Therefore, it is believed that further development of multi-fuel engines should be based on Common-Rail type systems [48, 55, 56].

Regulation of the quantity of supplied fuel, as well as start of ignition angle of the pilot dose should be assured. The publications of Stelmasiak [39, 40, 41, 42], Kowalewicz [19, 20], Mikulski, Wierzbicki and others [27, 28, 32, 52] have provided significant input into research on regulation parameters in multi-fuel engines. Moreover, there are also concepts for the introduction of new engine operation-control parameters. Many researchers report on the possibility to improve the efficiency of multi-fuel engines by controlling (throttling) quantity [41] and temperature [34, 35] of supplied air. In view of the reduction of the pilot dose to a minimum, the assurance of multi-fuel capability must bring about the necessity to take into consideration the composition of gaseous fuel in control algorithms [9, 16]. Recently, Liu and Fei [23] used neural networks technology to implement complex control algorithms in a dual-fuel engine.

1.3. Motivation and scope of the present research

The above analysis shows that in the upcoming years alternative gaseous fuels will significantly diversify all energy sectors. That is why research effort should be directed towards optimizing the chemical energy conversion processes for those fuels. Particular emphasis should be made on dual-fuel PCCI engines, as they show good potential in terms of fuel flexibility and clean and efficient combustion. As it has been discussed the concept of dual-fuel engines is still a new subject, and the development of optimum control algorithms requires further theoretical and experimental research. Main challenges to be conquered in terms of the combustion process control are: enabling efficient partial load operation with liquid dose minimisation, while keeping beneficial NOX and HC emissions levels and assuring stable transient operation. Even more complicated scientific problem remains to manage all of the mentioned issues for changing gaseous fuel composition.

The goal of the presented research was to discuss the possibility of utilizing alternative gaseous fuels in compression-ignition engines, using dual-fuel, gas-liquid operation strategy and contribute to research the problems in this topic with emphasis on fuel - flexible operation. The latest results of research on the impact of gas composition on combustion process had been presented, at the same illustrating the environmental benefits of using gaseous fuels. The Utilization of gaseous fuels with varying composition was illustrated systematically, starting with natural gas. The possibility of using low-calorie gases was shown by the impact of depleting natural gas with carbon dioxide. Industrial gases, such as syngas contain a large amount of hydrogen, carbon monoxide or higher hydrocarbons (ethane, propane). The possibility of fuelling CI engines with these gasses was presented by the influence of enriching natural gas with mentioned components.

The presented research cover engine dynamometer tests for different operating conditions with the analysis of the combustion process and detailed emission measurements discussion. The results of experimental studies were supplemented by simulation results, using mathematical models, developed by the authors for multi-fuel engines.

2. Engine test stand setup and methodology

The object used for the tests was a four-cylinder, turbocharged and intercooled CI engine, manufactured by Andoria-Mot, with a code mark ADCR. Basic technical parameters of the engine have been presented in Table 3. The engine has been equipped with a Common-Rail fuel injection system with piezoelectric injectors and was controlled by an EDC16C39 electronic controller with factory (optimized for diesel fuel operation) engine maps. The basic input signals introduced to the controller were engine speed and accelerator pedal position. The controller carries out two different fuel injection strategies, depending on the operation parameters. At low rpm and under small loads in the medium speed range, divided injection was performed. In the remaining range, a single fuel charge was injected.

Engine	**ADCR**
Type	diesel, 4-stroke, turbocharged with intercooler
Fuel injection	Common Rail fuel accumulator system
Engine layout	4 cylinder inline, vertical
Cylinder diameter / piston travel	94 / 95 mm
Piston displacement volume	2636 cm^3
Compression ratio	17,5 : 1
Rated Power / rotational speed	85 kW / 3700 rpm
Max. Torque / rotational speed	250 Nm / 1800-2200 rpm
Min. Idle rotational speed	750 rpm
Fuel consumption at torque peak	210 g/kWh
Injection system (Bosch)	accumulator injection system (Common Rail) CR2.0
Turbocharger	radial, with exhaust extraction valve
EGR system	pneumatic EGR valve with exhaust cooler

Table 3. Technical data of the ACDR engine.

The engine was installed on a test bed at the Department of Mechatronics and IT Education of the University of Warmia and Mazury in Olsztyn. The test stand in the configuration used in the research have been shown on Figure 1.

During the tests, engine speed and dynometer Torque were set as demanded values for each test point. The test bench automatics used dynamic dynometer operation and acceleration pedal operation to maintain the set values with accuracy of +-10 rpm and +- 5 Nm for speed and torque respectively. The cooling water and lube oil temperatures were kept at a constant 85/95 deg. C ± 1. After stabilization, steady state measurements were performed, including basic operation parameters of the engine, in-cylinder pressure, injector coil current, opacity and emissions. A time period of 120 s was set for data acquisition in every test run.

A piezoelectric pressure sensor (Type 6056A by Kistler), installed in the first cylinder through the heater plug adapter, was used for recording pressure signal. The sensor, combined with a Type 5018A charge amplifier, was connected via a DAQ card to a PC. The association of the pressure signal with respective rotation angle values was provided by an optical encoder mounted on the engine's crankshaft. The recording was performed every 1 CA deg., in the full range of the engine's work cycle.

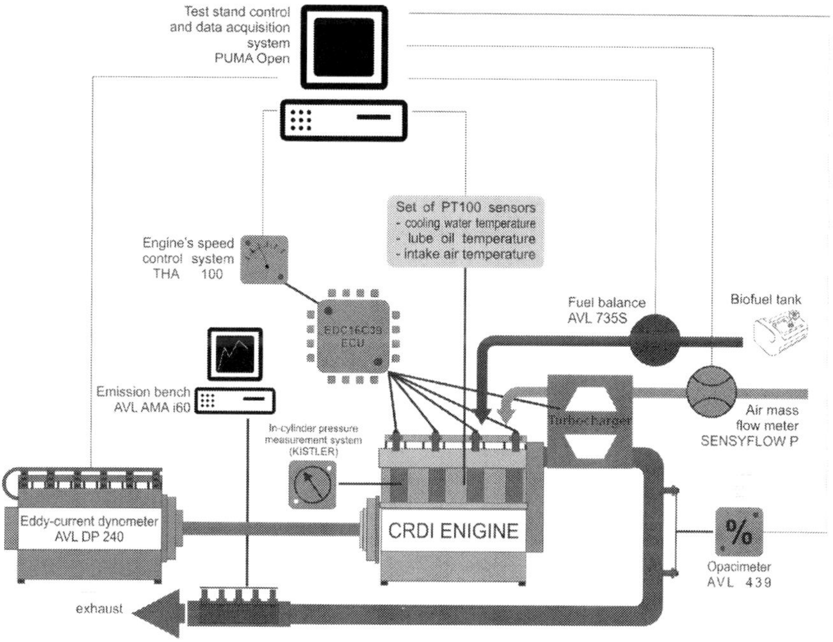

Figure 1. The scheme of the engine test bench used for research.

For each measurement point, after stabilization of the engine's operation parameters, pressure versus crank angle (α) was recorded for 100 cycles. Pressure measurement results were then cycle-averaged, giving pavg (α). Standard deviation was used to calculate the average error of pressure measurement for each crank angle Δpavg (α). The average relative error was then calculated:

$$\Delta p_{avg\_r}(\alpha) = \frac{\Delta p_{avg\_r}}{p_{avg\_r}} \cdot 100 \tag{1}$$

as an estimate of the engine's operation repeatability. Additionally, to determine the moment of injection, a current clamp was mounted on the injector of the indicated cylinder, which allowed recording the current changes on the injector coil. The injector coil current signal was analyzed in the same way as described for pressure signal and was used to determine the start of injection angle (SOI).

The mass of air aspired by the engine (G_{air}), fuel consumption (G_{fuel}) and air temperature at the inlet manifold (T_{air}) were recorded during the tests, and time averaged for each test run. Standard deviation was used to calculate the accuracy of the measurements results. A similar methodology was used to determine the measurement uncertainty of concentration levels of exhaust components.

With the use of the measured values, a number of parameters were calculated. The method of second pressure derivative analysis [32], was used in order to designate the start of combustion (SOC) and to calculate the ignition delay angle (α_{id}). Brake-specific fuel consumption (BSFC) and brake-fuel conversion efficiency (BFCE) were also calculated according to the following formulas:

$$BSFC = \frac{G_{fuel}}{P_e} \tag{2}$$

$$BFCE = \frac{3600}{BSFC \cdot Q_{fuel}} \tag{3}$$

where P_e – engine power, and Q_{fuel} – gross fuel heating value. For the calculated values, the measurement uncertainty was designated using the Kline and McClintock method [14]. The list of parameters recorded directly or indirectly during the test runs, along with maximum uncertainty for all measurement points, is presented in Table 4.

For the second phase of the research the ADCR engine described above was reconfigured, for the dual-fuel operation. The standard EDC16C39 engine controller was dismounted and replaced with a custom made liquid fuel injection control system. The system allowed for adjusting the injection angle and injection timing with the resolution of ±0,5 CA deg. and 1 μs respectively. During the experiments, a specific amount of gaseous fuel was fed to the intake manifold; its flow rate was regulated by mass flow controllers MASS-STREAM D-6371-DR. The separate flow regulator was used for CNG as the base fuel, and another controlled the CO_2 dilution of the gas-air mixture. Both liquid and gaseous fuel injection parameters where adjusted and monitored using a single PC (the same as for in-cylinder pressure signal recording). The scheme of the dual-fuel feeding system was presented on figure 2 [45, 53].

The per-cent energy fraction of liquid fuel (U_{fuel}) introduced to the engine at the given test point (set by engine speed and torque) in dual-fuel operation was designated as follows. The liquid fuel chemical energy introduced to the engine during mono-fuel operation was compared with the total chemical energy of liquid and gaseous fuel necessary to reach the same operating point during dual-fuel operation: the flow regulators

$$U_{fuel} = \frac{G_{fuel} \cdot Q_{fuel}\,|_{monofuel}}{G_{fuel} \cdot Q_{fuel} + G_{CNG} \cdot Q_{CNG}\,|_{dualfuel}} \cdot 100\% \tag{4}$$

Where Q_{fuel} is the heating value of the liquid fuel, G_{CNG} and Q_{CNG} stand for CNG consumption and heating value.

No	Parameter	Symbol	Measurement device	Uncertainty level	Unit
Directly measured					
1	Engine rotational speed	N	AVL DP 240	± 5	RPM
2	Torque	Te		± 2	Nm
3	Generated Power	Pe		± 0,2	kW
4	Air aspired to the engine	G_{air}	SENSYFLOW P	± 0,5	kg/h
5	Fuel consumption	G_{fuel}	AVL 735S	± 0,1	kg/h
6	Intake Air temperature	T_{air}	PT 100	± 0,2	K
7	Start of injection angle	SOI	Current clamp	± 0,5	CA deg.
8	Total hydrocarbons *	THC	AVL AMA i60	± 11	ppm
9	Total nitrogen oxides *	NO_X		± 19	ppm
10	Carbon monoxide *	CO		± 13	ppm
1	Carbon dioxide *	CO_2		± 1	ppt
10	Oxygen *	O_2		± 1	ppt
13	Opacity	EGO	AVL 439	± 0,9	%
Calculated					
14	Start of combustion angle	SOC	Second derivative of pressure analysis	± 1	CA deg.
15	Brake specific fuel consumption	BSFC	Equation (2)	± 10	g/kWh
16	Brake fuel conversion efficiency	BFCE	Equation (3)	± 1	%
*- concentration of the compound in the exhaust gases					

Table 4. List of parameters recorded directly and derived indirectly from calculation, along with achieved maximum uncertainty level.

In order to estimate the influence of higher gaseous hydrocarbons addition for dual-fuel engine operation, a simulation model by Mikulski has been used for preliminary research. A detailed description of the developed mathematical model, as well as the methodology of numeric calculations, can be found in another study by the authors [26]. Provided below is a concise summary of the key elements of the model.

The model implemented detailed chemical reaction kinetics of gaseous fuel – air mixture in a dynamic volume, zero-dimensional reactor describing the engine cylinder. The model includes the phases of compression, combustion and decompression in the chamber of a dual-fuel compression-ignition engine. It is assumed that at any point in time, the charge in the cylinder is a homogeneous mixture of air, natural gas, diesel fuel and exhaust fumes. Proportions of individual components vary in the stages of injection and combustion of combustible compo-

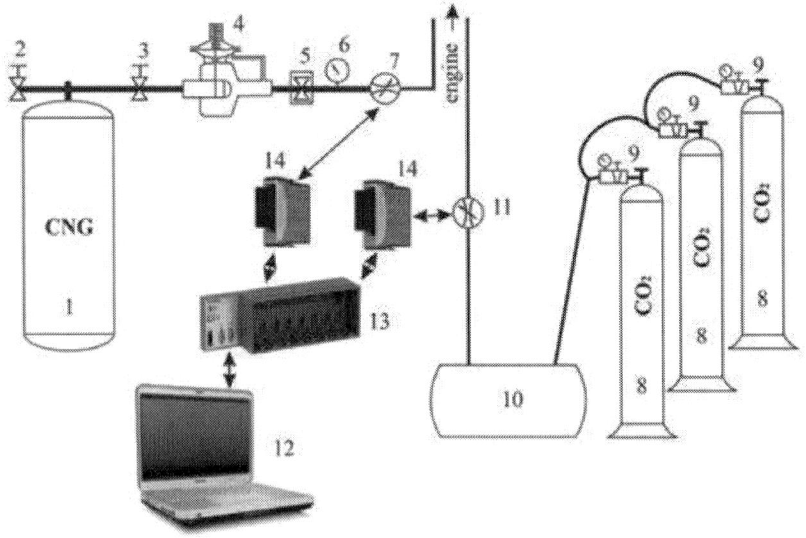

Figure 2. Block diagram of the gaseous fuel supply system: 1 - CNG cylinder, 2 - filling valve, 3 - cut-off valve, 4 - two-stage reducer, 5 - solenoid valve, 6 - pressure gauge, 7 - MasStream gas mass flow regulator, 8 - CO_2 cylinders, 9 - reducers with pressure gauges, 10 - expansion tank, 11 - mass flow regulator, 12 - PC, 13 - CompactRio programmable controller, 14 - i/o cards to control the flow regulators.

nents. The state of charge parameters in the cylinder was described with the use of the energy equation derived from the second law of thermodynamics and the equation of the ideal gas law.

The model included heat exchange with the walls of the combustion chamber as a sum of three streams passing through the cylinder wall and head and the bottom of the piston. During injection of liquid fuel, the thermodynamic parameters of the medium change. The impact of the fuel stream injected into the cylinder was simulated with the use of the authors' proprietary correlation, based on normal distribution [24].

The starting point of diesel fuel combustion is determined by the ignition delay period. The model used the equation proposed by Assanis et al. [1], due to the fact that it provided a more accurate description of the impact of the presence of gaseous fuel in the cylinder on the delay of auto-ignition of diesel fuel [32]. Diesel fuel works as an ignition inhibitor for the gaseous fuel - the ignition point is identical for both fuels. The potential ignition delay of gaseous fuel arises from the calculations of the applied gas combustion mechanism.

The diesel fuel combustion process in the model was simulated with the use of Wiebe's function, with the coefficient md = 0,4, obtained from verification tests.

This approach enables the analysis of the impact of the diesel fuel combustion process on the combustion process of the gaseous fraction. The model of gaseous fuel combustion was based on a one-step macro-reaction of direct oxidation of the main combustible components of the mixture: methane (CH_4), ethane (C_2H_6) and propane (C_3H_8), to carbon dioxide and water. The rate of heat release from the combustion of individual components is determined by the kinetics of the chemical reaction, with the following formula for global reaction rate:

$$\frac{d[C_iH_{2i+2}]}{dt} = A_n exp\left(-\frac{Ea_i}{\bar{R}T}\right)\cdot\left[C_iH_{2i+2}\right]^{a_i}\left[O_2\right]^{b_i} \quad (5)$$

The formulas in square brackets represent concentration levels of specific reagents. The solution of each of the differential equations (3) determines the course of changes of the number of moles of specific reagents (N_i (α)) and, at the same time, the heat release rate from the combustion of gaseous fuel:

$$\frac{dQ_g(\alpha)}{d\alpha} = \sum_{i=1}^{3}\frac{dQ_i(\alpha)}{d\alpha} = \sum_{i=1}^{3}H_i\frac{dN_i(\alpha)}{d\alpha} \quad (6)$$

The constant values found in equation (6) are summarized for specific gases in Table 5.

i		A	Ea	a	b	H_{mol}	H_{mas}
		[-]	[MJ/mol]	[-]	[-]	[MJ/mol]	[MJ/kg]
1	CH_4	$8,3\times10^6$	0,125	0,3	1,3	802,5	50,03
2	C_2H_6	$1,1\times10^{12}$	0,125	0,1	1,65	1423,7	47,3
3	C_3H_8	$8,6\times10^{11}$	0,125	0,1	1,65	2045,3	46,38

Table 5. Values of constants in Eq. (5).

The model was implemented in Matlab environment. The calculations were done, with the calculation step equivalent to 0,5 CA deg. The described model has been used to calculate: in-cylinder pressure, temperature and heat release rates for every CA deg.

3. Dual-fuel CNG-diesel engine test result

Numerous results available in publications dedicated to dual-fuel CI engines have proven that the effective use of such engines is possible within the maximum load range [26, 28, 40, 43].

It is related with the fact that with lower engine loads the air/gas mixture inside the combustion chamber is too lean and does not combust completely. It can be assumed that in this case only the part of the gaseous fuel, in close proximity to injected liquid fuel, is being burnt. The remaining part of the gaseous fuel (on the outskirt of the combustion chamber) does not combust, because the air/gas mixture is beneath its combustibility limits.

3.1. Standard engine controller operation

Tests on the effects of the use of dual-fuel CI engine were carried out using the engine described in the second paragraph. During the initial test phase the engine was run using the original controller - programmed for single-fuel operation, which performed different fuelling strategies depending on operation conditions. At lower rotational speeds the dose of injected liquid fuel was divided into the pilot and main doses, whereas at higher rotational speeds only one fuel dose was injected. Additionally, depending on the operating conditions, the controller changed fuel pressure in the accumulator rail and the angle of start of injection.

During this test phase a controlled dose of gaseous fuel was introduced into the inlet manifold. Inside the manifold and intake the gaseous fuel was mixed with air, creating homogenous mixture which was introduced to the combustion chamber. The liquid fuel dose was adjusted automatically by the controller, to ensure constant engine load output. Thus, as the gaseous fuel content in the total fuel dose was increased, the liquid fuel was reduced. By the change of the injected diesel fuel dose the controller simultaneously changed the start of injection angle.

Selected patterns of combustion chamber pressure registered for two rotational speed values, and recorded emission compositions for these measurements have been shown in figs. 3-5.

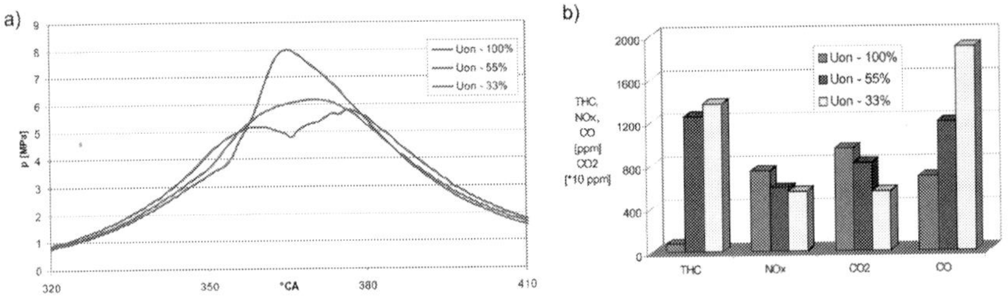

Figure 3. Combustion chamber pressure change patterns (a) and the concentration of toxic compounds emissions (b) at a rotational speed of 1500 rpm, Mo=150Nm for various shares of CNG in the fuel dose.

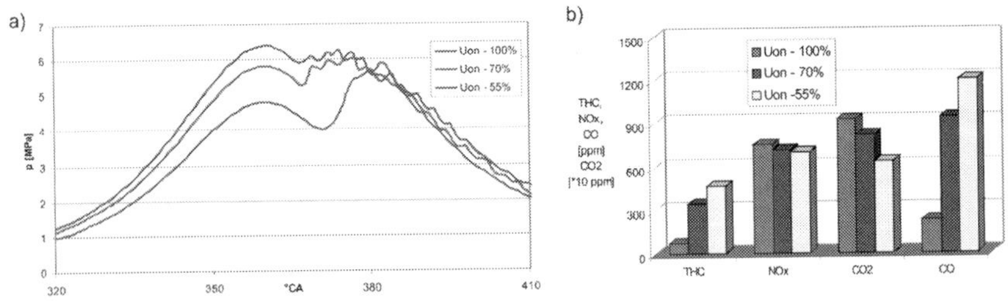

Figure 4. Combustion chamber pressure change patterns (a) and the concentration of toxic compounds emissions (b) at a rotational speed of 3000 rpm, Mo=150Nm for various shares of CNG in the fuel dose.

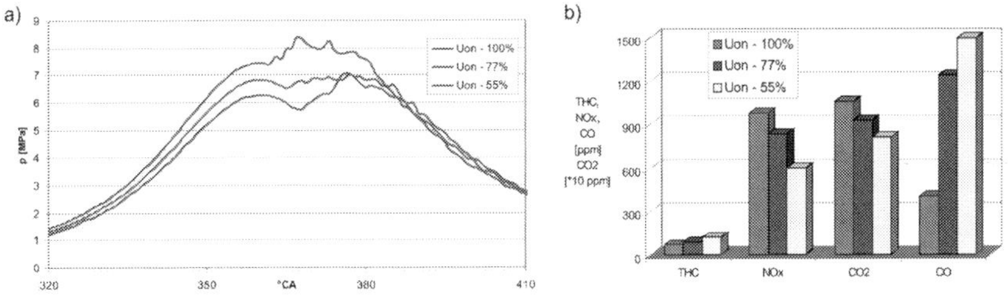

Figure 5. Combustion chamber pressure change patterns (a) and the concentration of toxic compounds emissions (b) at a rotational speed of 3000 rpm, Mo=200Nm for various shares of CNG in the fuel dose.

The presented results clearly indicate significant changes, both in the combustion process patterns and in emission levels. At lower rotational speeds (fig. 3), when a divided dose of liquid fuel is injected, the ignition of gaseous fuel is initiated by ignition of the pilot dose, which resulted in a sudden pressure rise in combustion chamber. At higher rotational speeds (figs. 4 and 5), when a single injection of liquid fuel was applied the rise of CNG content in the fuel dose delayed the pressure incensement, which was caused by delayed injection of liquid fuel. Thus, the combustion of gaseous fuel occurred after the piston reaches the TDC position. In every tested situation the increase of the CNG content resulted in a consequential increase of THC emissions, because of the incomplete combustion of methane. The increase of the gaseous part of the total fuel introduced into the combustion chamber also brings an increasement in CO emission levels, but NO_X and CO_2 emissions are slightly reduced.

With such fuelling strategy the gaseous fuel addition dose also significantly reduced the total engine efficiency, which must be related primarily to the incomplete combustion of the gaseous fuel. It must also be pointed out that the increase of gaseous fuel content in the fuel dose caused the engine to operate with more unevenness. With the content of gas fuel at the level of approx. 50% of the energetic value, engine operation became uneven and made it impossible to maintain constant engine load. It was assumed that the reason for this was the inappropriate injection of the diesel fuel.

3.2. Towards optimized dual-fuel control strategy

In order to define the effect of pilot dose parameters on the gas fuel combustion process in a CI engine, a special controller was used during the further phases of testing, allowing to control the liquid fuel dose parameters, such as injection pressure and start of injection. Control of diesel fuel pilot dose injection parameters, and above all the change of start of injection angle enabled a significant increase in the content of gaseous fuel in the fuelling dose, while maintaining smooth engine operation. Figs. 6-8 show the effect of the pilot dose injection advance angle on the combustion chamber pressure trace, as well as on the content of toxic compounds in exhaust gases, for different diesel fuel amounts.

The pressure patterns shown in the diagrams above indicate unequivocally that governance of the pilot dose injection does affect the air/gas mixture combustion process. Increasing the

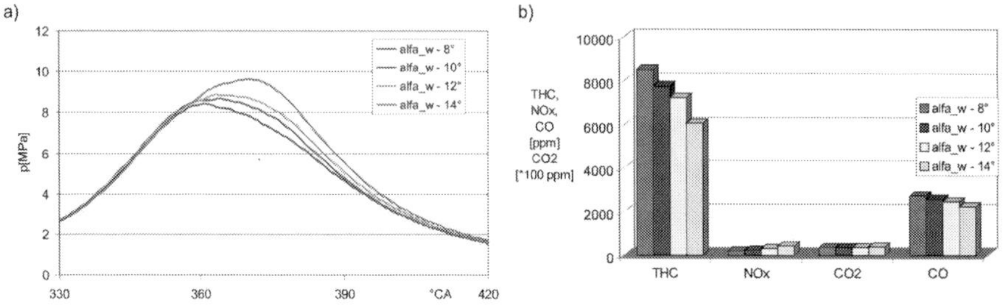

Figure 6. The effect of the pilot dose injection advance angle on the combustion chamber pressure change pattern (a) and toxic compound emission (b) at 3000 rpm, for a 24% content of diesel fuel in the fuelling dose.

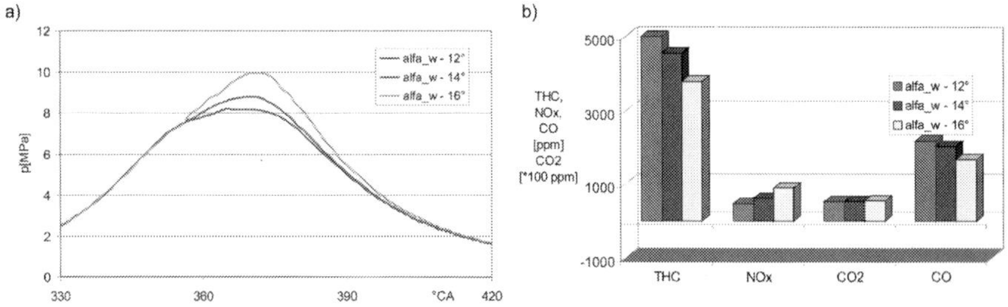

Figure 7. The effect of the pilot dose injection advance angle on the combustion chamber pressure change pattern (a) and toxic compound emission (b) at 3000 rpm, for a 20% content of diesel fuel in the fuelling dose.

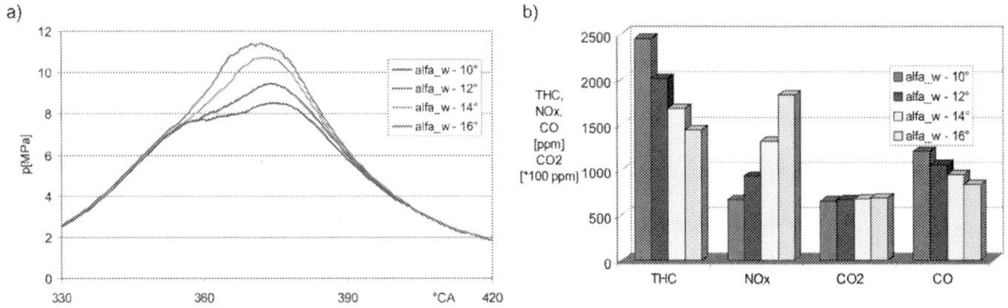

Figure 8. The effect of the pilot dose injection advance angle on the combustion chamber pressure change pattern (a) and toxic compound emission (b) at 3000 rpm, for a 15% content of diesel fuel in the fuelling dose.

injection advance angle of the diesel fuel pilot dose accelerates the combustion of gas fuel, which leads to the increase in maximum combustion chamber pressure and better combustion of gaseous fuel. Changing the injection advance angle does also affect the amount of toxic compounds in emissions. Increasing this angle causes a significant reduction of THC and CO emissions, which should contribute to better combustion of gaseous fuel. Unfortunately, alongside the increase of the injection advance angle, the amount of NOx in emissions also rises as a result of the rise in combustion chamber temperature due to better combustion of methane.

The change of the injection advance angle of the diesel fuel pilot dose also affects engine performance. Fig. 9a illustrates the effect of the pilot dose injection advance angle on the torque value, while fig. 9b illustrates the effect of this angle on engine efficiency for different values of diesel fuel content in the fuelling dose.

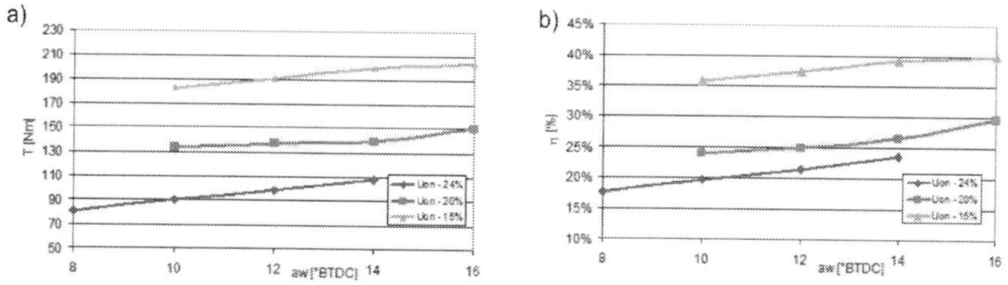

Figure 9. Effect of the injection advance angle of the pilot dose on the value of torque and general engine efficiency at 3000 rpm, and various values of diesel fuel content.

The results presented above allow us to conclude that the effective use of gaseous fuels is possible in the high engine load range. This is mainly due to the fact that at large engine load values the produced gas/air mix has better parameters that assure better combustion of gas fuel. In the case of a lesser gas fuel content the mixture is too lean and its combustion is not complete and too slow, and thus it does not guarantee the achievement of high engine efficiency.

It must be noted, however, that the presence of gaseous fuel in the fuelling dose leads to higher maximum pressure values in the engine, which in turn causes higher loads imposed on the engine's piston-crank mechanism. When the content of gaseous fuel rises, the content of THC and CO is reduced, this contributes to the reduction of the air surplus coefficient value. However, the emission value of NO_X increases due to higher combustion temperatures.

3.3. The effects of diluting CNG with CO_2 on engine performance and emissions

Due to the recent rise in interest in renewable fuels, including biogas, in the next testing phase several measurements were made in order to determine the effect of adding CO_2 to gaseous fuel.

Figs. 10 and 11 illustrate registered combustion chamber pressure change patterns and the content of toxic compounds emissions for different values of CO_2 addition in gaseous fuel. The illustrated combustion chamber pressure patterns clearly indicate that as the content of CO_2 in gas fuel rises, the ignition of gaseous fuel is delayed, which, in constant pilot dose injection conditions, leads to a reduction in combustion chamber pressure and, in effect, to the reduction of engine efficiency. Fig. 12 illustrates the effect of the content of CO_2 in gas fuel on torque and general engines' efficiency values, with constant diesel pilot dose injection parameters.

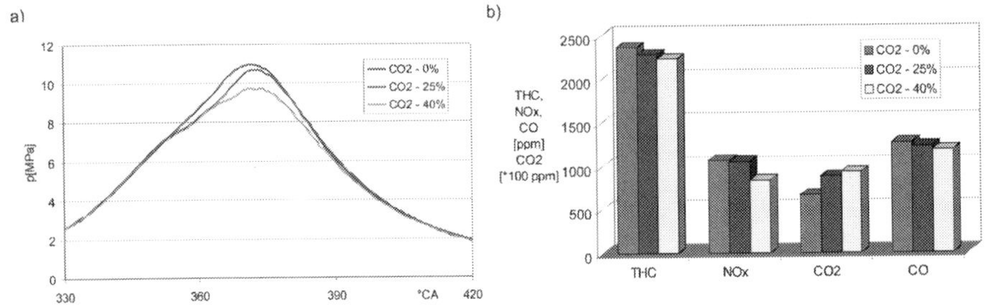

Figure 10. Combustion chamber pressure change patterns (a) and patterns of the presence of toxic compounds (b) at 3000 rpm, with diesel fuel content at 22.5%, for various CO_2 content levels.

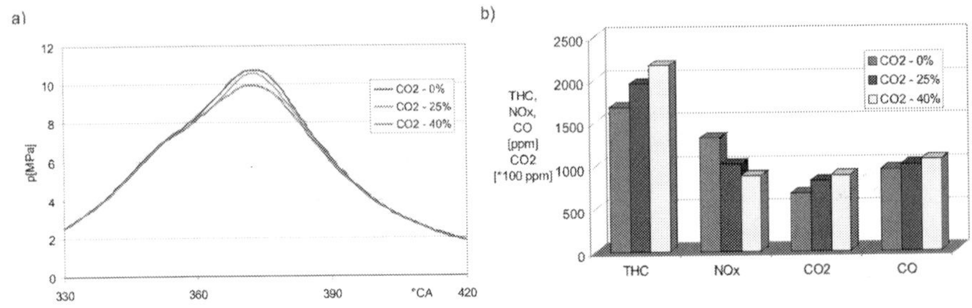

Figure 11. Combustion chamber pressure change patterns (a) and patterns of the presence of toxic compounds (b) at 3000 rpm, with diesel fuel content at 15%, for various CO_2 content levels.

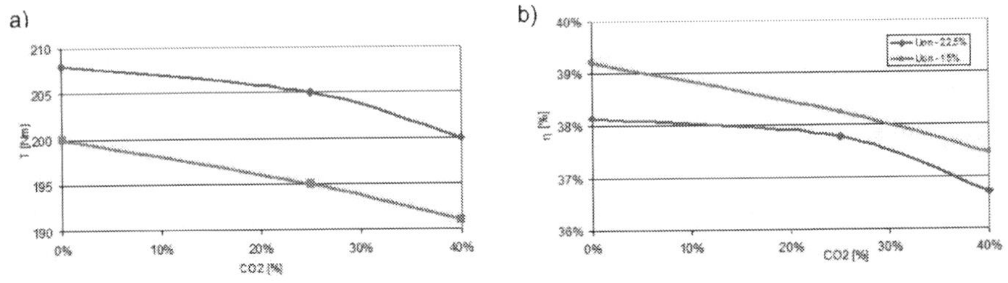

Figure 12. Effect of content value of CO_2 in gas fuel on the value of torque and general engine efficiency at 3000 rpm and various values of diesel fuel content.

During the performance tests on dual-fuel CI engines, a series of experiments was also carried out in order to determine the effect of liquid fuel injection pressure on engine's performance. Fig. 13 presents combustion chamber pressure change patterns and patterns of toxic emissions for different diesel fuel pilot dose injection pressure values. Fig. 14 illustrates the effect of pilot dose injection pressure values on the gained torque and general engine efficiency.

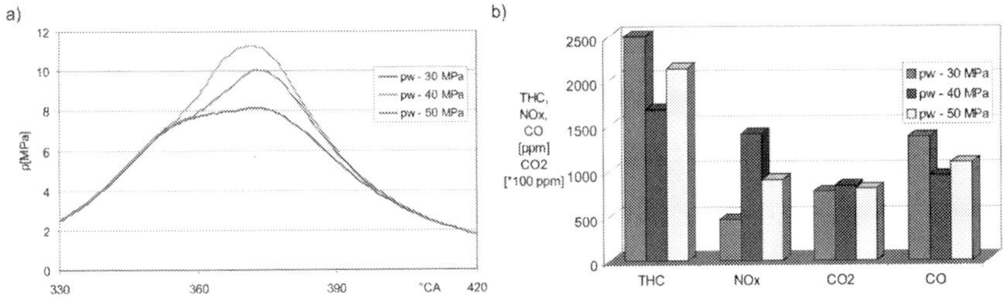

Figure 13. Combustion chamber pressure change patterns and toxic compounds patterns at 3000 rpm, with diesel fuel content at 15%, content of CO_2 in gaseous fuel at 25%, for various pilot dose injection pressure levels.

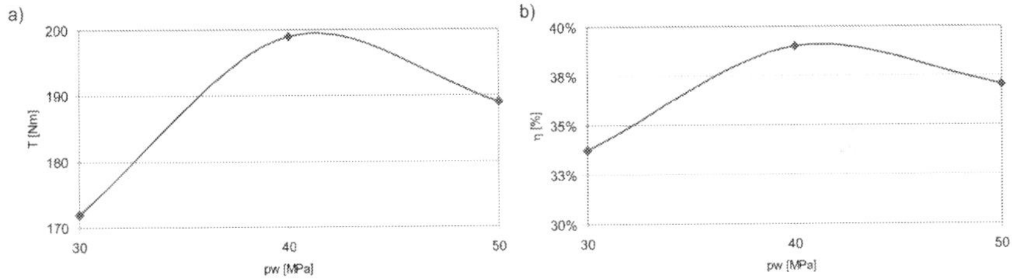

Figure 14. Engine torque and general engine efficiency patterns at 3000 rpm, with diesel fuel content at 15%, content of CO_2 in gaseous fuel at 25%, for various pilot dose injection pressure levels.

The test results presented above suggest that if the pilot dose injection pressure is too low, it causes a significant reduction in general engine efficiency, which is most probably connected with insufficient fuel spraying, which results in the insufficient speed of flame propagation within the combustion chamber and, in effect, gas fuel is not subject to complete combustion. If the injection pressure is too high, on the other hand, it causes surplus fuel spraying, which is also detrimental to the combustion process. The obtained results are compliant with the results described in [40, 43, 53],who note that the optimum results of the operational efficiency of a dual-fuel engine can be obtained with limited pilot dose injection pressure values.

3.4. The effects of C_2H_6 and C_3H_8 enrichment on engine performance and emissions

In order to show the influence of addition of higher hydrocarbons, simulation tests were performed. As the starting point for the study, the operating parameters of an ADCR engine were used, at a rotational speed of 3400 rpm and break torque of 150 Nm. The correctness of the model for these parameters was verified for a broad range of gaseous fuel proportions [25].

Calculations were done for a constant share of gaseous fuel of 70% of the total energy supplied in the fuel. The angle of start of injection of liquid fuel was common for all measuring points and amounted to 6 CA deg. before TDC. The start of ignition was calculated by the model, providing for the impact of gas composition. Similarly, the amount of injected liquid fuel (2,2 kg/h) and its combustion duration (23 CA deg.) were constant for all simulations. It was assumed that the entire volume of gas supplied to the cylinder would participate in the reaction, and the combustion of the gaseous fuel was complete. The composition of gaseous fuel, determined by the share of individual components, was the simulation parameter. The tests assumed a constant level of gas and air consumption.

We shall illustrate the effect of using higher hydrocarbons as gaseous fuel by studying large substitutions of ethane and propane (10%, 30% 80%), for which the effect is clearly visible. Since both additions have different influence on the diesel fuel combustion, also an effect of addition even amounts of those gasses have been briefly described. The results of model simulation for all those cases was shown on figs. 15-17.

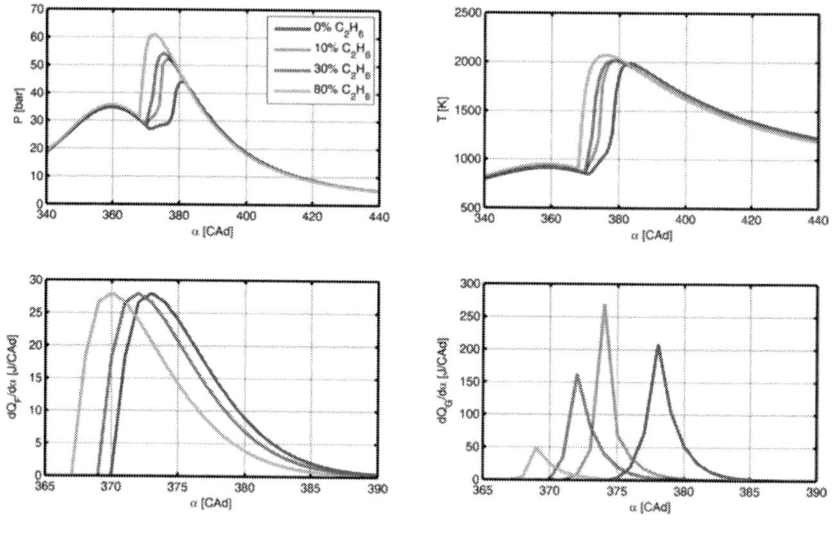

Figure 15. Pressure and temperature values in the cylinder and heat release rate from the combustion of diesel fuel and gas. Different methane content with respect to ethane (C_2H_6).

For most gaseous fuels, a delay of start of combustion is observed with the increase of their concentration levels in the charge [16]. In the case of ethane addition to gaseous fuel, it has been observed that its presence in the engine cylinder reduces the auto-ignition delay of diesel fuel (fig. 15). This reverse effect of ethane has also been observed by other researchers. Research [16] has indicated that with highly reactive gases, such as hydrogen or ethane, the presence of the gas may reduce the ignition delay period. This tendency arises, for the most part, from the high activity of the gases in pre-flame reactions and the creation of auto-ignition locations even before the injection of diesel fuel.

Fig. 15 indicates that even a small addition of ethane also significantly accelerates the ignition of gaseous fuel. In the case of using pure methane, gas would ignite ca.5 CA deg. after the

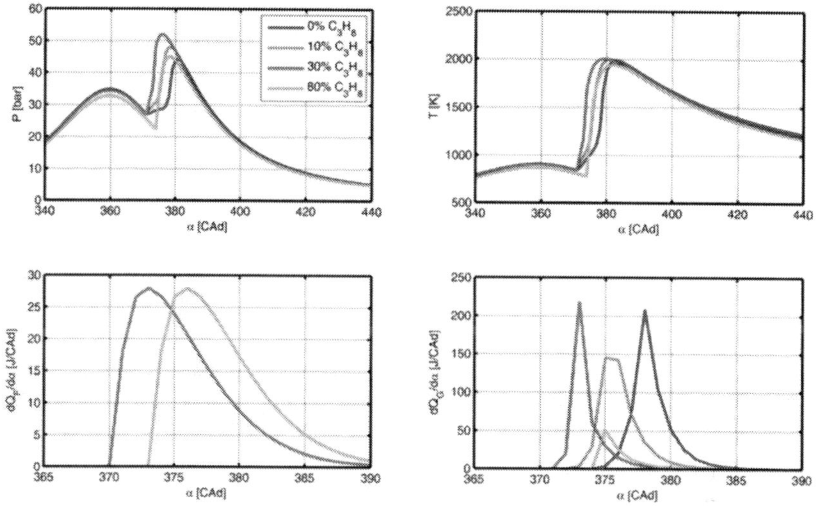

Figure 16. Pressure and temperature values in the cylinder and rate of heat release rate from the combustion of diesel fuel and gas. Different methane content with respect to propane (C_3H_8).

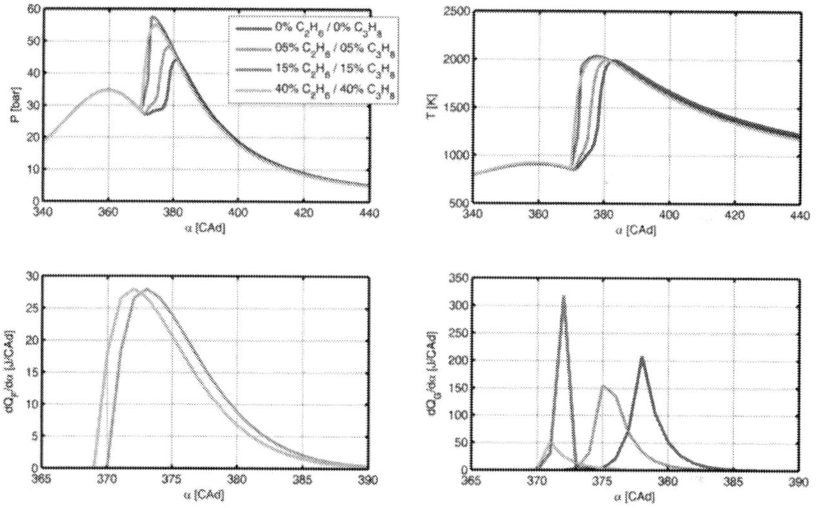

Figure 17. Pressure and temperature values in the cylinder, and rate of heat release from combustion of diesel fuel and gas. Different methane content with respect to ethane (C_2H_6) and propane (C_3H_8).

ignition of liquid fuel. On the other hand, an addition of 20% of ethane already caused the gaseous fuel to ignite together with the initial dose ignition.

The addition of propane results in a minor increase of the diesel fuel ignition delay, as shown by the data in fig 16. A slight extension of the auto-ignition delay has been observed repeatedly for increasing the percentage weight of ethane in the analysed fuel. The delta between the extreme samples (100% methane and 20% methane / 80% propane) exceeded 3 CA deg. This had significant impact on the pressure and temperature in the cylinder – fig. 16.

The length of gaseous fuel combustion was slightly higher for samples containing up to 20% additions of both C_2H_6 and C_3H_8, as compared to the base samples - marked as 100% CH_4 (fig. 15 and 16). For higher concentrations of ethane and propane, a slight increase of combustion rate has been observed, correlated to the increase of maximum heat release rate. In each case, the earlier ignition of gaseous fuel would generate a more rapid increase of pressure and temperature, combined with the increase of maximum combustion pressure and temperature. An increase of maximum pressure by 33% for even a small (10%) addition of C_2H_6 significantly improves the performance of the engine, powered by ethane-enhanced gaseous fuel, as compared to base methane fuel. The observed increase of combustion temperature (by 5% on average) with even minor enhancements with ethane may, at the same time, increase the emissions of NO_X.

Important data on the combustion process in the engine powered by the mixture of diesel fuel, methane and ethane was provided by the analysis of the heat release rate from combustion of both gaseous combustible components This analysis indicated that earlier ignition of gaseous fuel is caused by earlier ignition of ethane, which requires less energy to initiate combustion. Rapid heat emission from combustion of ethane also accelerates the ignition of methane, which, however, has always had a slower ignition than ethane. With identical percentage weights (50%), ethane burns longer than methane, at the same time generating smaller heat increments, despite its higher calorific value. An increase of the length of burn of gaseous fuel as a whole, observed for additions of up to 20% of methane, arises in equal proportions from the distribution in time of combustion of individual gaseous fractions (ethane ignites quicker than methane) and from the slower rate of the ethane oxidation reaction.

Similarly to ethane addition, the pressure and temperature of combustion of the methane-propane mixture was determined mainly by the kinetics of combustion of gaseous fuel. Identically to ethane, propane accelerated the ignition of the gaseous mixture by lowering the activation energy. The quicker ignition of propane enriched gas mixture generated a higher maximum temperature and pressure than for pure diesel/methane; however, due to the increase of ignition delay caused by increasing the concentration levels of propane, the obtained increments were smaller than for ethane. This effect also explains why, despite the higher calorific value of propane, the sample with the highest concentration level of the gas (80% propane) generated smaller increments of pressure and temperature than 10% propane.

In the case of adding equal volumes of ethane and propane to methane, the ignition point of the diesel fuel does not visibly change (fig. 17). This means that the accelerating effect of ethane and decelerating effect of propane appear to cancel each other out. The combustion process is determined by the moment of ignition of gaseous fuel, which occurs for all additions, together with liquid fuel ignition (fig. 17). Still ethane seems to have stronger inhibiting effect on diesel fuel auto-ignition, compared to the ignition delay increase caused by propane.

4. Conclusions

It is clear that methane – based gaseous fuels, especially off renewable origin, will have an important role in global energy share in the nearest future. It was proven that modern CI engines operating in dual – fuel PCCI mode are a good solution for effective and clean gaseous fuel utilization.

The material discussed proves that dual-fuel technology is already mature enough to be used at steady state operation – for energy generation for example. We have showed that basic regulation strategies can help minimizing diesel consumption to 10% of total energy introduced with the fuel. The engine efficiency can be maintained at the same level as for standard diesel engine and in some cases slight efficiency improvement can be reached. This is combined with lower CO and CO_2 emission and can be maintained for different gaseous fuel composition including low calorific gases with high CO2 content.

The major drawbacks that remain are: higher NOx and THC emmision and poor performance at low loads – to be dealt with more sophisticated control algorithms. Also successful transient control is a challenge – due to the fact that mutual interactions of gaseous and liquid fuels haven't been sufficiently understood. Future implementation should focus on dealing with dose issues - discovering new control concepts for the engine itself, combined with methane aftertreatment technology development.

Also the posiboilitie of introducing a multi-fuel engine, which is able to operate on different gaseous fuels, have been shown. Gaseous fule composition changes have a significant effect on the combustion process.

Also it was proven that real progress in the field of multi-fuel RCCI, PCCI and HCCI concepts is possible with proper, control oriented simulation tools as support. Development of dose tools should be therefore an equally important goal as development of new engine concepts itself.

Author details

Mikulski Maciej[1,2*], Wierzbicki Sławomir[2], Ambrosewicz-Walacik Marta[2], Duda Kamil[2] and Piętak Andrzej[2]

*Address all correspondence to: maciej.mikulski@uwm.edu.pl

1 The Netherlands Organization for Applied Scientific Research (TNO), Department Powertrains, Helmond, Netherland

2 University of Warmia and Mazury in Olsztyn, Faculty of Technical Sciences, Mechatronics and IT Education, Olsztyn, Poland

References

[1] Assanis D.N., Filipi Z.S., Fiveland S.B., Syrimis M.. A predictive ignition delay correlation under steady-state and transient operation of a direct injection diesel engine. J. Eng. Gas Turbines Power. 2003;125 (2):450-457. DOI: 10.1115/1.1563238.

[2] Bekdemir C., Baert R., Willems F., Somers B.. Towards control-oriented modeling of Natural Gas-Diesel RCCI Combustion. SAE Technical Paper. 2015;DOI: 10.4271/2015-0101745.

[3] BP p.l.c., editor. BP statistical review of world energy. London, United Kingdom: 2012.

[4] BP p.l.c., editor. BP energy outlook 2030. London, United Kingdom:2013.

[5] Brosseau J. D., Zajic J. E.. Continuous microbial production of hydrogen gas. Int. J. Hydrogen Energy. 1982;7:623-628.

[6] Carlucci A., De Risi A., Loforgia D., Naccarato F.. Experimental investigation and combustion analysis of a direct injection dual-fuel diesel-natural gas engine. Energy. 2008;33:256-263.

[7] Cebula J.. Biogas purification by sorption techniques. ACEE Journal. 2009;2.

[8] Curran S., Prikhodko V., Cho K., Sluder C. et al.. In-cylinder fuel blending of gasoline/diesel for improved efficiency and lowest possible emissions on a multi-cylinder Light-Duty Diesel Engine. SAE Technical Paper. 2012;DOI: 10.4271/2010-01-2206.

[9] Czerwiński J., Comte P.. Influences of gas quality on a Natural Gas Engine. In: Proceedings of the International Scientific Conference "Development and Design of ice Engines and Quality of Fuels to Meet Future Levels"; 2001; Kraków, Poland.

[10] Fuerhaper A. Unger E., Piock W. F., Fraidl G.. The new AVL CSI engine - HCCI operation on a multi cylinder gasoline engine. SAE Technical Paper. 2004;:2004-01-0551.

[11] Hairuddin A. A., Wandel A. P., Yusaf T.. Hydrogen and natural gas comparison in diesel HCCI engines – a review. In: Proceding of the 2010 South Region Engineering Conference (SREC 2010); 2010.

[12] Han, X., Xie, K., Zheng, M., De Ojeda, W.. Ignition control of gasoline-diesel dual fuel combustion. SAE Technical Paper. 2012;:2012-01-1972. DOI: 10.4271/2012-01-1972.

[13] Holladay J. D., Hu J., King D. L., Wang Y.. An overview of hydrogen production technologies. Catalysis Today. 2009;139 (4).

[14] Holman, J. P.. Experimental methods for engineers. 6th ed. McGraw-Hill:1994.

[15] Inagaki, K., Fuyoto, T., Nishikawa, K., Nakakita, K. et al.. Dual-Fuel PCI Combustion Controlled by In-Cylinder Stratification of Ignitability. SAE Technical Paper. 2006;:2006-01-0028. DOI: 10.4271/2006-01-0028.

[16] Karim G.A.. Combustion in gas fueled compression: ignition engines of the dual fuel type. J. Eng. Gas Turbines Power. 2003;125 (3):827-836. DOI: 10.1115/1.1581894.

[17] Karner D., Francfort J.. High-Percentage Hydrogen/CNG Blend Ford F-150 operating summary. Idaho National Engineering and Environmental Laboratory Bechtel BWXT Idaho, LLC. 2003;:INEEL/EXT-03-00007.

[18] Kong S. C., Reitz R. D.. Use of detailed chemical kinetics to study HCCI engine combustion with consideration of turbulent mixing effects. Journal of Engineering for Gas Turbines and Power. 2002;124:702-712.

[19] Kowalewicz A., Wojtyniak M.. Natural gas engines – problems and challenges. Journal of KONES. 2007;14.

[20] Kowalewicz A.. Adaptacja silnika wysokoprężnego do zasilania gazem ziemnym (*Adaptation of a diesel engine to natural gas supply*). Czasopismo Techniczne. 2008;:zeszyt 7-M.

[21] Lechwacka M.. Technologie uszlachetniania biogazu do jakości gazu ziemnego (*Technologies of processing biogas to natural gas quality*). Czysta energia. 2009;12.

[22] Leermakers C., Somers L., Johansson B.. Combustion phasing controllability with dual fuel injection timings. SAE Technical Paper. 2012;:2012-01-1575. DOI: 10.4271/2012-01-1575.

[23] Liu Z., Fei S.. Study of CNG /diesel dual fuel engine's emissions by means of RBF neural network. Journal of Zhejiang University Science. 2003.

[24] Mikulski M.. The problems of multi-fuel engines modelling in the context of systems operational parameters estimation – Part I - fuel injection. Zeszyty Naukowe Politechniki Warszawskiej. 2011;5 (86):113-124.

[25] Mikulski M.. Budowa dwufazowego modelu spalania w wielopaliwowym silniku o zapłonie samoczynnym (*Construction of a two-phase combustion model in a multifuel, compression-ignition engine*) [thesis]. Wydział Nauk Technicznych, UWM Olsztyn, Poland:2014.

[26] Mikulski M., Wierzbicki S., Piętak A.. The multi-phase, zero-dimensional, computational model of a multi-fuel CI engine fueled with gaseous fuel with divided injection of liquid fuel. Eksploatacja i Niezawodnosc – Maintenance and Reliability. 2015;17 (1):42-48.

[27] Mikulski M., Wierzbicki S., Piętak A.. Numerical studies on controlling gaseous fuel combustion by managing the combustion process of diesel pilot dose in a dual-fuel engine. Chemical and Process Engineering - Inzynieria Chemiczna i Procesowa. 2015;36 (2):225-238. DOI: 10.1515/cpe-2015-0015.

[28] Mikulski M., Wierzbicki S., Śmieja M., Matijošius J.. Effect of CNG in a fuel dose on compression-ignition engine's combustion process. Transport. 2015;30 (2):162-171. DOI: 10.3846/16484142.2015.1045938.

[29] Mohan B., Yang W., Chou S. K.. Fuel injection strategies for performance improvement and emissions reduction in compression ignition engines - A review. Renewable and Sustainable Energy Reviews. 2013;28:664-676. DOI: 10.1016/j.rser.2013.08.051.

[30] Nitkiewicz S., Duda K.. Uzdatnianie metanu pofermentacyjnego (*Purification of post fermentation methane*). Combustion Engines. 2011;:2011-SC-163.

[31] Ormerod J. G., Ormerod K. S., Gest H.. Light dependent utilization of organic compounds and photoproduction of molecular hydrogen by photosynthetic bacteria, relationship with nitrogen metabolism. Arch Biochem Biophys. 1961;94:449-456.

[32] Piętak A., Mikulski M.. On the modeling of pilot dose ignition delay in a dual-fuel, self ignition engine. Combustion Engines. 2011;3 (146):94-102.

[33] Premier A.. HCCI could cut fuel consumtion by 15%. Advanced Materials and Processes. 2007;11:27-44.

[34] Saito H., Kawabata Y., Sakurai T.. Study on the lean burn gas engine ignited by pilot fuel injection (part 3). Annual Technical Report Digest. 1999;9.

[35] Saito H., Kawabata Y., Sakurai T.. Study on the lean burn gas engine ignited by pilot fuel injection (part 4). Annual Technical Report Digest. 2000;10.

[36] Saravan N., Nagarajan G., Narayanasamy S.. An experimental investigation on di diesel engine with hydrogen fuel. Renewable Energy. 2008;87:3591-3599.

[37] Sasikala K., Ramana Ch. V., Raghuveer P. R.. Photoproduction of hydrogen from the waste water of a distillery by Rhodobacter sphaeroides O.U.001. Int J Hydrogen Energy. 1992;17:23-27.

[38] Skorek J., Kalina J.. Gazowe układy kogeneracyjne (*Gaseous cogeneration systems*). Warszawa, Poland: WNT; 2005.

[39] Stelmasiak Z.. Wpływ wielkości dawki inicjującej oleju napędowego na parametry pracy dwupaliwowego silnika zasilanego olejem napędowym i gazem ziemnym (*Effect of the magnitude the diesel fuel initial dose on the work parameters dual fuel engine fed diesel fuel and natural gas*). Journal of KONES. 2000;8.

[40] Stelmasiak Z. Studium procesu spalania gazu w dwupaliwowym silniku o zapłonie samoczynnym zasilanym gazem ziemnym i olejem napędowym (*Study of the combustion process of gaseous fuel in dual fuel compression ignition engine fueled with natural gas and diesel oil*). Wydawnictwo ATH; 2003.

[41] Stelmasiak Z., Larish J., Gilowski T., Matyjasik M.. Możliwość poprawy składu mieszaniny gazowej przez dławienie powietrza przy częściowych obciążeniach silnika

dwupaliwowego (*The possibility of quality improvement of gas-air mixture at part load of dual fuel engine*). Archiwum Motoryzacji. 2007.

[42] Stelmasiak Z.. The combustion controlling in the dual fuel CI engine by pilot dose devision. Combustion Engines. 2011;3.

[43] Stelmasiak Z.. Dwupaliwowe silniki o zapłonie samoczynnym (*Dual fuel compression ignition engines*). Radom, Poland: ITE; 2013.

[44] Surygała J.. Wodór jako paliwo (*Hydrogen as a fuel*). Warszawa, Poland: Wydawnictwo Naukowo-Techniczne; 2008.

[45] Śmieja M., Wierzbicki S.. The concept of an integrated laboratory control system for a dual - fuel diesel engine. Journal of KONES. 2012;19 (3):451-458.

[46] Tanisho S.. Biological hydrogen production by fermentation. In: Proceedings of the Second International Conference on New Energy Systems & Conv.; 1995; Istanbul.

[47] Tomita E.. Dual fuel HCCI combustion – high octane and high cetane number fuels. International Journal of Engine Research. 2005;6:453-463.

[48] Wadman B.. Dual fuel conversion of Cat D399 Engine. Diesel & Gas Turbine Worldwide. 1999.

[49] Wang Z., Shuai S.J., Wang J. X., Tian G.H.. A computional study of direct injection gasoline HCCI engine with secondary injection. Fuel. 2006;85 (12-13):1831-1841.

[50] Wierzbicki S.. Biogas as a fuel for diesel engines. Journal of KONES. 2012;19:477-482.

[51] Wierzbicki S.. Analysis of the effect of the chemical composition of low calorific gaseous fuels on workload concentration in an engine's combustion chamber. Journal of Polish CIMAC. 2013;8:89-96.

[52] Wierzbicki S.. Laboratory control and measurement system of a dual-fuel compression ignition combustion engine operating in a cogeneration system. Solid State Phenomena. 2014;210:200-205.

[53] Wierzbicki S., Mikulski M., Śmieja M.. Effect of CO_2 content in CNG on the combustion process in a dual-fuel compression ignition engine. Combustion Engines. 2015;162 (3):91-101.

[54] Yao M., Zheng Z., Liu H.. Progress and recent trends in homogeneous charge compression ignition (HCCI) engines. Progress in Energy and Combustion Science. 2009;35:398-437.

[55] Zeltner R.. Engine management for dual fuel power plants. Diesel & Gas Turbine Worldwide. 1998.

[56] Zhang Y., Liu X., Yang Q., Han X., Cheng Ch.. The studies of electronically controlled CNG system for dual fuel engines. SAE Technical Paper. 2001;:2001-01-0145. DOI: 10.4271/2001-01-0145.

[57] Zhong S., Wyszynski M., Megaritis A., Yap D., Xu H.. Experimental investigation into HCCI combustion using gasoline and diesel blended fuels. SAE Technical Paper. 2005;:2005-01-3733. DOI: 10.4271/2005-01-3733.

Chapter 5

Liquefied Petroleum Gas (LPG) as a Fuel for Internal Combustion Engines

Maciej Paczuski, Marcin Marchwiany, Ryszard Puławski, Andrzej Pankowski, Kamil Kurpiel and Marcin Przedlacki

Additional information is available at the end of the chapter

http://dx.doi.org/10.5772/61736

Abstract

Liquefied hydrocarbon gases, containing mostly the mixture of C3-C4 hydrocarbons, in many countries serve as an important source of energy, also for internal combustion engines. On a world scale, at present, about 270 million tonnes of this fuel is produced, of which motorization consumes 26 million tonnes (~10%). Poland is one of these countries, where LPG is a popular engine fuel. Annual consumption of so called autogas is about 1.7 million tonnes with increasing tendency. Almost 3 millions of cars with bifuel engines, mostly spark ignition ones, are driven on Polish roads. An extensive logistical infrastructure accompanies the development of the use of this type of fuel in Poland. There are numerous production companies in Poland working to meet the needs of this branch of industry, as well as a modern service base. Intensive research and development works in this field are carried out. The exceptional advantages of LPG as an alternative engine fuel have been highly appreciated in Poland. These advantages result mainly from the simplicity of its production. Relatively low investment costs and energy consumption are needed to produce high-quality fuel by degasolining of natural gas or crude oil stabilization. LPG is produced as a byproduct in numerous petroleum refining processes and its quality is relatively close to the engine fuel requirements. The source of aliphatic hydrocarbons falling into LPG range can also be the processes of biomass conversion and in the near future, fermentation processes. Environmental advantages of LPG are primarily very low emission of almost all toxic components of engine exhaust gases compared to gasoline or diesel fuel. In this chapter the results of research works aimed at the improvementof exploitation properties, logistics and LPG combustion processes in spark ignitionengines are presented. In this area, anti-corrosion additives for LPG of very high effectiveness were obtained and tested. Also, with the help of additives, the problem of water separating from LPG during storage and transport in tankers has been solved. A multifunctional additive was obtained giving LPG adequate lubricity and detergent properties. The influence of water on the process of LPG combustion in a gasoline fuelled engine as well ways to supply water to the combustion zone are being studied. A mathematical model is being elaborated, precisely describing correlations between density and chemical composition of LPG as well as changes of density as a function of fuel temperature. In the realization of these plans LPG producing companies, research and development teams as well as producers of automobile LPG systems are involved. Successful completion of research works and design works will result in elimination of scarce disadvantages of LPG as an engine fuel, preserving at the same time all its numerous advantages as an alternative fuel for the future - the new era source of energy.

Keywords: Liquefied petroleum gases (LPG), corrosion, , water, additives, combustion, density modeling

1. Introduction

The mixture of liquefied hydrocarbon gases C_3-C_4 (propane and butane), called colloquially liquefied gas or LPG, is a particular energy carrier, counted among the group of alternative fuels. LPG has more than 1000 different uses, including applications in industry, civil engineering, communal economy, agriculture, households, and transport. Because of simplified logistics of transport ensuring supply diversification, availability of sources, and most of all environmental aspects, LPG exhibits high dynamics of production and consumption; the global production of this fuel comes close to 280 million tonnes.

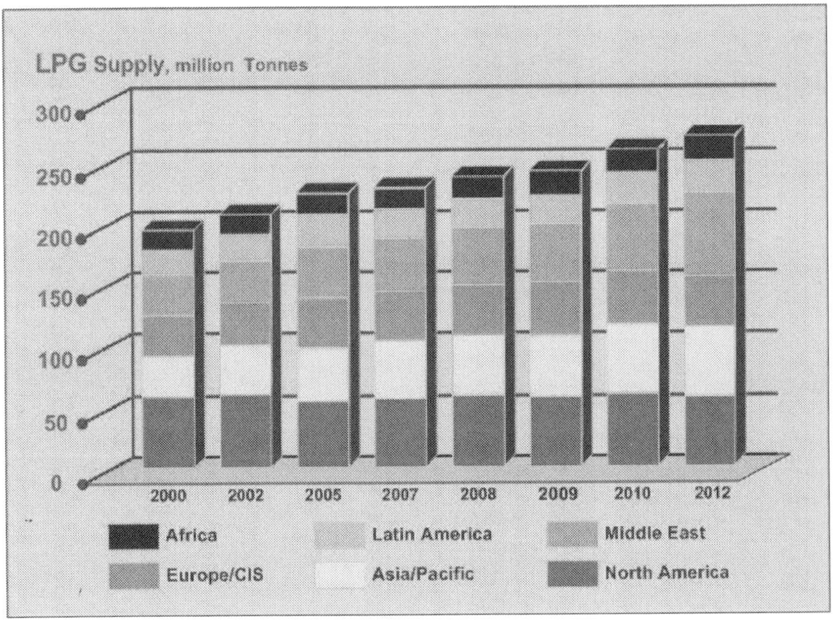

Figure 1. Global consumption of LPG [1]

Only a small part of global LPG production, about 10%, is used to fuel internal combustion engines.

Liquefied gas is produced from three main sources:

- As a byproduct from degasolining of natural gas,
- As a byproduct from stabilization of petroleum in crude oil mining site,
- As a byproduct from petroleum refining in a refinery.

Natural gas contains mainly methane, but also other substances, among them heavier hydrocarbons, including C_3 and C_4. Such gas is called "wet" and its preparation for transport requires the removal of LPG fraction – degasolining. Additional amounts of LPG are obtained during stabilization of crude oil at a mining site, being a part of petroleum preparation for the transport. It is estimated that globally about 60% of LPG is obtained that way [2]. Purified "liquid gas" contains mainly propane, butane, and isobutane in different proportions, but it does not contain unsaturated hydrocarbons. This makes it a fuel of highest quality and best chemical stability.

The other 40% of LPG is obtained during petroleum refining in a refinery. Depending on the kind of crude oil, it contains 1–4% of LPG fraction. The source of saturated hydrocarbons, being components of liquefied gas in a refinery, is distillation operations and numerous types of hydrotreating. Most fractions obtained from distillation of high-sulfur feedstock are hydrotreated. The wide spread of hydrotreating processes in the manufacturing of refinery products was caused by strict environmental regulations. In this way the feedstocks for the isomerization of light gasoline fractions and catalytic reforming of gasolines are prepared. Hydrodesulphurization of catalytic cracking feedstock is being introduced to refinery schemes. Hydrotreating is used to purify the components of jet fuel, diesel fuel, light heating oil, base lubricating oils and petroleum waxes. Deep petroleum refining means also the use of destructive hydrogen processes, hydrocracking of vacuum distillates and residua. As an effect of reduction with hydrogen of sulfur, nitrogen, and oxygen compounds, hydrocarbons are formed, including C_3-C_4 – Figure 2.

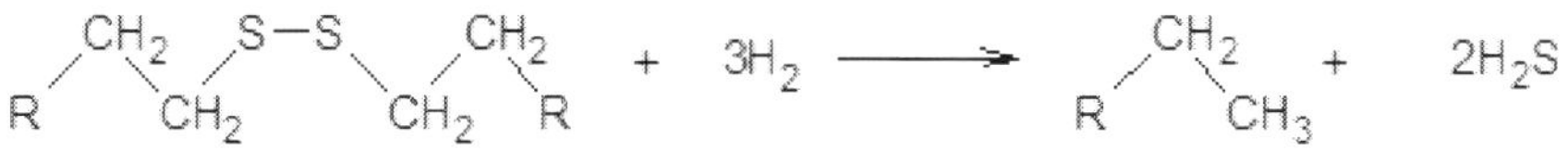

Figure 2. An example of hydrogenation of alkyl disulfide

LPG fractions obtained from thermal destructive processes, such as cracking or pyrolysis have slightly different chemical nature. They contain olefins and dienes, valuable chemicals, which can be found in LPG fractions used as motor and heating fuels.

Summarizing, it can be said that LPG is obtained as a byproduct from petroleum and natural gas refining, which causes that it must be a cheap product. Its improvement for the use as a motor fuel requires most often the reduction of sulfur content to the level required by environmental regulations.

The technology of autogas production can be much more complicated in situations when a fuel of normalized quality has to be produced from various components of varying chemical composition. Additional technological and logistical complication in autogas production can be a variation in properties of this fuel depending on the season of the year and climatic conditions.

The quality requirements for LPG in Poland are the same as in the EN 589+A1:2012 standard. The only differences are related to, so called, winter parameters, which each country defines adequately to its climatic conditions. LPG in Poland, since 2007, is subject to the same national fuel monitoring and control system, as other fuels, that is, gasolines and diesel fuels. The regulation governing this system is one of the most strict in Europe. The quality monitoring of motor fuels is supervised by the Office of Competition and Consumer Protection (UOKiK) and is carried out at each stage of logistical chain, starting from the producer, through shipment, storage, transport, and finally fuel stations. For many years, the results of LPG sample analyses have been very good. In 2014, only 1.5% of the LPG samples collected at fuel stations did not fully fulfill the normative requirements for these fuels. No deviations from required properties were found for samples collected from the tanks belonging to the producers and wholesalers. Only great effort undertaken by the autogas industry in the area of optimization of procedures and high investment in quality control systems led to obtaining such results.

What is characteristic for LPG quality requirements is that they do not contain its hydrocarbon composition. On the other hand, the producers cannot use any mixtures they choose, because the need to use particular chemical compositions is caused by the requirements for other key properties influencing the engine operation. These parameters include:

- Motor octane number
- Relative vapor pressure at 40°C
- Temperature at which the vapor pressure is not lower than 150kPa

The adequate value of motor octane number (≥89) of a fuel is required for correct and economical combustion in an engine. Indirectly, the standard ensures that LPG as a motor fuel has a research octane number, used to characterize gasolines, on a level higher than 95, often above 100.

It has to be noted that autogas evaporates in 100% in a cylinder, which cannot be said about gasolines and particularly about diesel fuels. Full evaporation is a factor deciding that LPG, despite having potentially lower calorific value than gasoline, does not have to be consumed in higher quantities.

Easily available sources of autogas components, increasing production of natural gas from unconventional sources, accompanied by C_3-C_4 hydrocarbons as well as reports relating to production of propane from biomass conversion indicate that this fuel has good perspectives for the increase of supply and further spread of its use.

http://dx.doi.org/10.5772/61736

LPG is an exceptional fuel, which is indicated by not only its availability but also by production costs. The comparison of costs or energy needed to produce modern gasolines or diesel fuels indicates that the production of LPG is much cheaper [3, 4].

LPG is rightly treated as an environment friendly fuel, not only because of production costs, but also because of social and ecological profits. A very dangerous and socially costly component of gases emitted from internal combustion engines is particulate matter (PM). The comparison of health costs of motorization is presented in Table 1 [5].

Component of exhaust gases	Health costs, EUR/t
$PM_{2,5}$	160.000
SO_2	10.000
NO_2	15.700
VOC	700
CO	20

Table 1. Evaluation of health costs related to the emission of components of exhaust gases

Engines fueled with autogas emit incomparably lower amounts of these components, which causes that total social costs of using different motor fuels are lowest for LPG – Fig. 3.

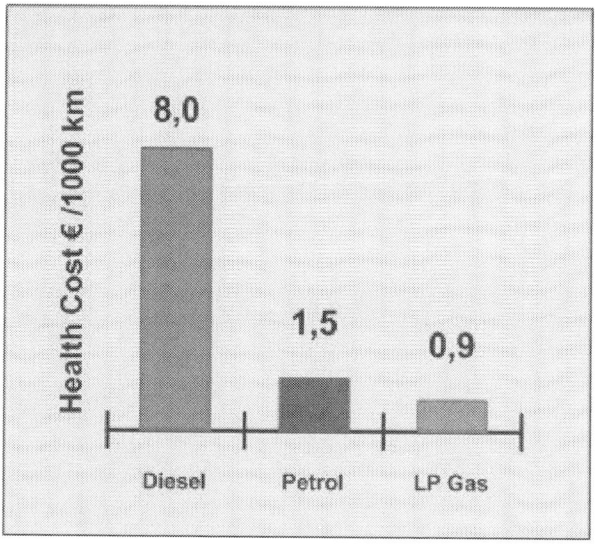

Figure 3. Health costs of using different motor fuels

Higher calorific value of autogas, compared to gasoline and diesel fuel, along with higher hydrogen to carbon ratio, leads to emission of lower amounts of carbon dioxide (73.6 g CO_2/MJ while for gasoline it is 85.8 CO_2/MJ, and for diesel fuel - 87.4 g CO_2/MJ.

The use of LPG as an alternative motor fuel needs relatively small and not expensive adaptations in spark-ignited engines. Automobile producers play a bigger and bigger role in the popularization of autogas. They noticed the economical and ecological profits from the use of such engines and more often offer automobiles with factory installed autogas systems. It creates a positive image of autogas as a fuel environmentally friendly, economical, and fully safe.

2. Autogas in Poland

For the energy policy, in the conditions of sustainable development, total costs of motor fuels, well-to-wheel, are very important, if the number of vehicles with internal combustion engines and the amount of fuel consumed by them are taken into account. Poland, in this aspect, is in a very good situation because of a vast number of bi-fuel vehicles and high participation of autogas in motor fuel consumption. More and more common treating of LPG as an alternative fuel is profitable for its wider use, both in transport and in other areas.

The beginnings of autogas industry in Poland reach the last decade of the twentieth century, when first legal LPG automobile installations appeared. Intensive development of this sector of the economy occurred at the turn of the century. The amount of autogas sold in 1996 was 250000 tons, in 2003 it was more than 1 million tonnes, and in recent years it got stabilized at the level of 1.5–1.65 million tonnes.

In Poland, the number of vehicles equipped with LPG installations constantly increases. After more than 20 years of autogas history in Poland, it is currently more than 2.85 million. It means that almost one in seven cars on Polish roads is fueled with LPG. It is estimated that globally the number of such cars is higher than 25 million, and Poland, along with South Korea, Turkey, Italy, and Australia, is a country with the highest population of bi-fuel cars (gasoline-LPG).

This fuel is easily available in Poland, because the number of autogas stations has reached 5500 and since 1998 it increased three-fold. Since a few years the increase of network stations offering autogas can be noticed. Novelization of regulations, allowing unassisted autogas fueling, fosters this process.

Many firms and companies import and distribute LPG in Poland, having well developed infrastructure, means of transport, and most of all qualified personnel. A fragmented structure of the import favors competition on the market and full elasticity in choosing the direction from which the liquefied gas is imported.

Polish LPG market in 2013 (2200000 tonnes) is composed of 17.7% of local production (Orlen Gaz, Lotos Paliwa, PGNiG) and imports from the Russian Federation (over 50%), Kazakhstan (over 25.1%), and Belarus (11.8%). Poland also exports LPG [6]. The structure of the LPG market has not been changed for years. The main role is played by autogas (almost 75% of the market). The second place is taken by 11 kg tanks (over 13%), used in 5.4 million households. About 12% of the market encompasses the segment of industrial and household LPG storage tanks.

During the last 20 years numerous companies were formed, producing LPG for the needs of local market and developing export to countries located on all continents. Basing on incomplete data, it can be estimated that 13 local producers of automobile LPG systems produce annually about 2 million such installations. These companies employ from several dozens to about 500 employers each. The characteristic feature of these companies is usually a wide profile of production, designing new elements of these installations and control systems. It is connected to having their own design and technology supply base, and some of these companies have research and development centers, cooperating with scientific and research institutes.

An example of such a company can be AC SA from Bialystok, noted on the GPW stock exchange. It is a leading company in Poland and a notable global producer of automobile LPG/CNG systems marked STAG. On the Polish automobile market, AC SA is present for 27 years, employs over 500 people and at present it is the largest company operating in the area of automobile gas installations in the country. As the company's evaluations indicate, it has over 50% share in the Polish market. The AC SA products reach over 40 countries all over the world. In 2011, the export made 63% of the AC SA total revenue. The key foreign recipients of AC SA include Germany, Russia, Ukraine, Turkey, and Thailand [7].

Qualified personnel, modern machinery park, efficient management, and marketing allow the expansion on foreign markets, including contracts with automobile producers for the delivery of "factory installed" LPG installations for new vehicles.

The development of the industry and increasing quality requirements of the customers, as well as the development of new engine technologies, initiated the process of formation of networks of garages authorized by the producers, with monitored service quality, elongation and widening of warranty range for offered LPG systems.

High safety and quality level of installations produced in Poland is an effect of very strong internal and foreign competition as well as fully functioning this branch in open market conditions.

The presence of foreign companies, mainly Dutch and Italian, representing highest global level, increases the competition and favors the technological development of Polish producers.

Over 20 local companies and a few foreign ones produce parts for the producers and service of LPG installations. Among them are both small and young companies as well as medium and big firms with long tradition. Five companies produce LPG tanks, adjusted to the varied needs of recipients. Over 1 thousand types and models of LPG tanks are produced. Local makers produce their own reducer-evaporators of liquefied autogas, injectors, pressure hoses, plastic elements, gas, and liquid phase filters. All producers are also exporters, cooperating with customers on all continents. One of the producers opened his representative office on very quickly developing the US market. The predominant directions of export for local producers are Eastern European countries as well as "old European Union", South America and Far East.

The dynamically developing LPG market in Poland induced the creation of companies designing and realizing investments in the area of construction of fueling stations, terminals, and fuel bases. These companies include design groups, firms installing and servicing IT systems for LPG industry, as well as telemetry systems, companies legalizing equipment and tanks. In total there are at least 10 companies of various size in this area.

In Poland there are about 1550 active companies installing and servicing automobile LPG systems [8]. Most of these workshops have the authorization of a given installation producer, maintaining contacts and training meetings with installation makers. Well-trained installers often help the customers to choose optimal solutions, passing their remarks and notifications to the producers. In this way they are cooperating in the creation of new solutions and improvement of technologies.

The innovative character of the LPG industry is realized as a natural need to remain on local market as well as to expand its actions abroad.

High popularity to use LPG as a fuel for spark-ignited engines caused that it started to be used in diesel engines. Diesel gas systems for engines fueled with diesel fuel and LPG give fuel consumption reductions from 10% to 20%. More efficient combustion of diesel fuel increases the engine power, reduces the emission of carbon (II) oxide and particulate matter, while it does not reduce the lifetime of an engine.

3. Quality issues of LPG as a motor fuel

Autogas, having many advantages, exhibits also several disadvantages compared with modern gasolines. The standard requirements do not allow to introduce additives to LPG, which in gasolines provide many key properties, such as cleaning of fuel supply system, reducing the need for octane requirement increase (ORI), thermal and thermo-oxidative stability, as well as anti-corrosion properties.

The aim of the research works described in this chapter is:

- Improvement of the quality of LPG in terms of corrosive activity against copper
- Increase of the accuracy of LPG analysis, in particular the determination of total sulfur content and corrosion against copper, through a new way of LPG sampling for analysis
- Improvement of LPG logistics taking into account the water content in the fuel as well as the influence of the chemical composition and temperature of the fuel on its density
- Improvement of LPG combustion process in an engine by introducing water into the combustion zone

3.1. Corrosion

One of the most important quality parameters of autogas is lack of corrosion activity against copper. In practice, however, the situation can happen that the corrosivity of this fuel appears in the logistics chain. Good quality LPG shipped via railway tankers and/or stored in tanks can dissolve chemical compounds that can cause the corrosion of copper plates. There is much evidence that the cause of this corrosion threat is water which stays at the bottom of tanks and contains various micro-contaminations. These substances in water undergo i.a. hydrolysis reactions, generating corrosive compounds, such as hydrogen sulfide.

In analytical practice, there are controversies related to the corrosivity of LPG. Using high quality instruments does not guarantee the repeatability of results. It appears that exceptional attention should be paid to the procedures of fuel sample collections.

3.2. Collection of LPG samples

Collection and transfer of representative sample of a material has key influence on the evaluation of its quality parameters. Proper collection of a representative sample has particular importance because of the increase of quality requirements for LPG, development of analytical techniques, and higher experience of the laboratory personnel. Often the operation of collection and preparation of the sample is the most neuralgic point, influencing the correctness of quality parameter evaluation for this batch of a product.

The issue of proper transfer of analytical sample from a product batch was undertaken during comparative studies organized by the Automotive Industry Institute (PIMOT) in 2013 [9]. These studies were carried out in almost ideal conditions, where all participants were prepared for the collection of samples, preparing all equipment according to every procedure in order to obtain maximum purity of the samples. The cleanliness of the sample collection equipment and the ability to maintain it in daily use, using present solutions and with high frequency of sample collection is very hard to achieve. As practice shows, often for random comparative studies on samples, collected during everyday operations, some of the results differ significantly, particularly for parameters such as total sulfur content and corrosivity against copper. It can be related to the difficulty with maintaining high cleanliness of the equipment.

The differences in the assessment of obtained results are often the subject of disputes between LPG importers and producers and controlling authorities, monitoring the quality parameters included in standards for LPG fuel. Presently used solutions and their effects in the form of differences in analytical results of the same LPG fuel are often cause for fines for the companies taking part in LPG trade. The authority verifying the quality is the State Trade Inspection (PIH), exerting regulations in the name of the Office of Competition and Consumer Protection (UOKiK) and Energy Regulatory Office, and in the case of finding that the fuel does not fulfill quality standards, it imposes fines.

In Poland, an innovative solution was used, enabling precise evaluation of the quality and classification of fuels, not only protecting final recipients but also the interests of individual entrepreneurs, at the same time stimulating the further development of this industry.

Available literature does not discuss this issue in the aspect of possible analytical problems, except for the current PN-EN ISO 4257 standard, related to the collection of LPG samples. Taking into account the development of the technology and problems that are met in LPG analytics, a solution has been elaborated which can eliminate this problem. This solution allows collection and transfer of a representative sample from product batch and then its proper evaluation [10].

New technical solution, related to the demountable sampler of specially prepared internal surface, has been subject of profound comparative studies. In these studies, the results of LPG quality parameter measurements using two different types of samplers and ways of cleaning them were compared. During these studies, the influence of sampler design and the way of its preparation for sample collection as well as its cleanliness on the representativeness of quality assessment was evaluated. Studies were carried out in the area of the evaluation of the quality of collected LPG samples and microscopic scanning of internal surfaces of the samplers.

Non-damaging and damaging tests on the samplers of two different designs were carried out in order to obtain information on the influence of sampler design and its internal surface on the processes occurring in the sampler and the representativeness of the sample. Standard non-demountable sampler, made of chromium-nickel 316L steel, was compared to a demountable sampler made from chromium-nickel 316Ti steel. Additionally, in a new demountable sampler, the internal surface was prepared by mechanical polishing, finished with etching and chemical electropolishing. The surface prepared this way was passivated with oxides. Figure 4 presents the two types of samplers, along with their cross-sections.

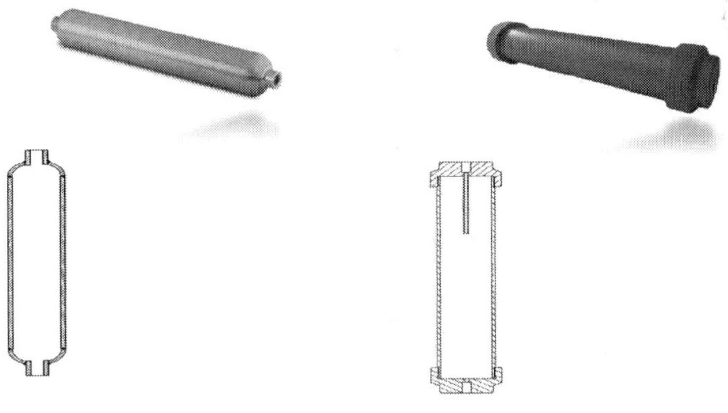

Figure 4. Traditional, non-demountable sampler [left], demountable sampler [right]

In the case of a demountable sampler, as a new solution, using the processes of etching and chemical electropolishing was aimed at the decrease of surface microporosity, which reduces the threat of aggregation of deposits and chemicals on the surface and makes it easier to clean. Operation applied for the preparation of the internal surface of the sampler uncover homogeneous, clean metal surface, close the micropores, precluding chemical contaminations to adhere to the surface. Electropolishing facilitates further passivation of the surface. The role of the passive surface is protection from adhering chemical compounds to the internal surface of the sample, which can accumulate and influence the correctness of results obtained during long term use of the sample.

Non-damaging, microscopic scans of an internal surface of the sampler showed the surface structure in both types of solutions. The surface of a non-demountable sampler (Figure 5), in places where no additional finishing was used, is very uneven, has noticeable scratches, and sharp edges. On such surfaces, contamination can very easily adsorb and accumulate, having influence on the results of LPG analysis collected with this sampler. High surface roughness makes it hard to clean, which causes the possibility that secondary reactions in the cylinder can occur and the quality parameters of collected samples can change. In the case of a demountable sampler (Figure 6), where additional finishing was applied to the internal surface, it is smooth, the pores of the material are closed, and the structure is homogeneous. Surface prepared in this way impedes the deposits and contaminations to adhere and accumulate on it, increasing the representativeness of collected samples. Lack of possibility to accumulate contaminations simplifies cleaning, which is more efficient. Thus, the sampler is cleaner and provides better representativeness of obtained analytical results.

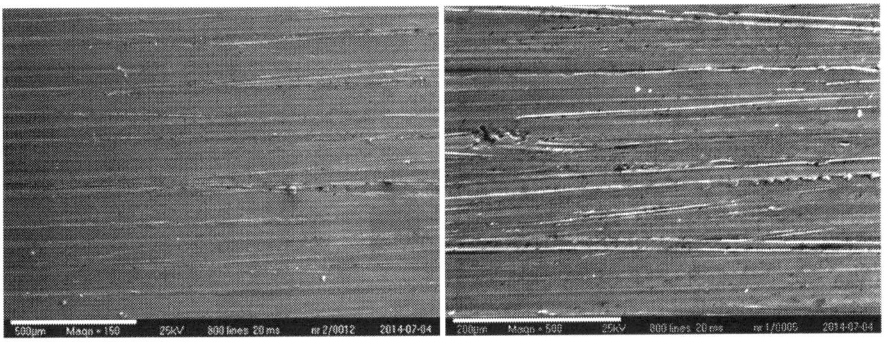

Figure 5. Internal surface of a non-demountable sampler (150x magnification [left], 500x magnification [right])

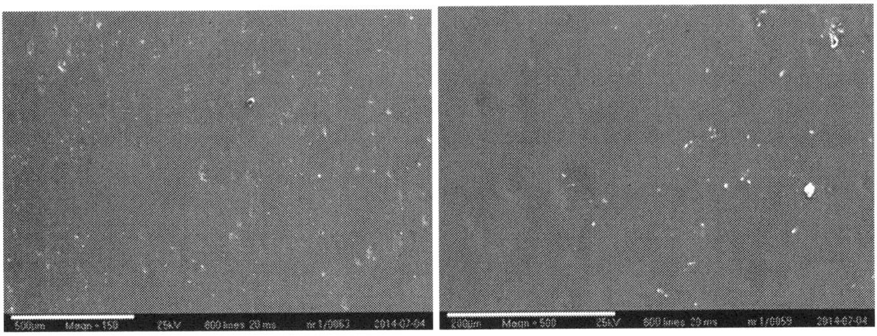

Figure 6. Internal surface of a demountable sampler (150x magnification [left], 500x magnification [right])

Also destructive tests have been carried out that demonstrated the differences in both types of samplers. Each sampler was used to collect 250 samples of LPG. Samplers were cleaned, according to the PN ISO 4257 standard, before each collection. After a series of tests, both samplers were cut along their axis in order to analyze the internal surface of a cylinder. In the case of a new, demountable sampler, no contamination was observed. In the traditional, non-demountable sampler, a dark deposit was observed on its internal surface. Analysis of the deposit indicated the presence of such elements as Na, K, Mg, Ca, Al, V, Ni, Co, Mn, Fe, and S. The most undesirable element in this group is sulfur. The analyzed deposit contained 0.72% m/m of this element. Sulfur in the deposit occurs mainly in the form of inorganic compounds, which in favorable conditions can undergo secondary reactions, forming hydrogen sulfide and sulfur.

Samples collected using both types of samplers under identical conditions were analyzed qualitatively in the same way. In different periods of time, during 3 years, 2113 analyses were made according to the standard procedures. For the parameters such as corrosive activity against copper and total sulfur content, the deviations of the results for some samples had decisive importance for proper qualitative classification of the fuel. Some analytical results for samples collected with a traditional sampler indicated qualitative inconformity of tested fuel with the requirements of the regulations. Biggest deviations were noted for the parameter "corrosive action against copper" – about 8% deviation of the results – and for "total sulfur content," where the differences quantifiability reached 11%. It has to be noted that the increase in discrepancy of obtained results with the time of use of samplers was observed. This fact can confirm the thesis that secondary reactions occur, having direct influence on the analytical result and consequently on the representativeness of the samples.

The results of comparative studies in operation of the two types of LPG samplers, a traditional one and a newly designed construction, indicate that the new solution has significant advantages, particularly for the objective assessment of LPG quality in the area of total sulfur content and corrosive action against copper. These parameters are at the same time the most frequent causes of exceeding the LPG quality standards.

The value of the new solution is confirmed by awarding it a patent [11] and registering it as an European registered design.

This technical solution and studies on its application was twice distinguished and awarded in the "Napęd Nowej Ery" (New Era Drive) competition. It has been used since 2010 with very good results, not only for collecting LPG samples but also for analyses of other technical gases of very high purity.

Strict qualitative requirements and the specifics of LPG transport and storage require in some situations the use of particular additives, in order to fulfill standard parameters or for efficient and safe logistics.

3.3. Anti-corrosion additive

In Poland the problem of LPG corrosivity was solved a few years ago by elaboration and introduction of an anti-corrosion additive of a very high effectiveness [12]. The application of this additive in an LPG exhibiting the corrosivity level higher than 1, up to 4, causes reduction of the corrosivity class to level 1a. The additive contains a composition of slightly basic nitrogen compounds in high concentration in an appropriate mixture of solvents which provide good miscibility of active substances with liquefied C_3-C_4 hydrocarbons, practically in all temperatures. Nitric bases have the ability to neutralize acidic, corrosive components of LPG, hydrogen sulfide, and mercaptans. In turn, the affinity to metals of surfactants, contained in the additive, causes the creation of an adsorptive layer on the surfaces of copper and also steel. This layer forms a protective barrier and also improves the lubricity of the fuel containing the additive. The lubricity action of the additive, confirmed in High Frequency Reciprocating Rig (HFRR) tests, causes the reduction of average friction coefficient from 0.591 to 0.257, and 10-fold reduction of wear (by 89.3%). A significant advantage of this product has to be underlined – it is mixed with LPG in a 1:8000 (v/v) ratio and practically it does not increase a standard parameter residue after evaporation (oil residue). Experiences collected during several years of use of this additive by Polish LPG producers and importers, as well as observations made by foreign customers, confirm the high efficiency of the additive.

4. Water in the fuel

LPG exhibits very limited solubility of water but despite this, technological operations carried out at increased temperatures and then cooling during transport or storage cause the separation of water. As a result, at the bottom of tanks and railway tankers, almost always there is a layer of water. The unloading of LPG, particularly during autumn or winter, causes in such situations the blockage of unloading valves with ice. This situation is particularly difficult for automobile tankers supplying autogas fueling stations. Similar problems can occur in the distributors of LPG while fueling cars with water-contaminated fuel. The practice of LPG logistics, as well as European standards, recommend in such situations adding to the tanks and tankers small amounts of methanol. It is an accepted solution, but having some disadvantages. Methanol mixes with water in all proportions, lowering its freezing point, but it does not mix with the fuel. In this way, the volume of water phase in the tank increases, which can cause its increased transfer to further volumes in the logistical chain. The original solution used in Poland is based on the addition of a special additive to the water-contaminated fuel, which increases the "solubility" of water in LPG [13]. This additive contains surfactants exhibiting the ability to solubilize water in the fuel. The affinity between water molecules and the surfactant increases with the decrease of temperature. This effect wholly protects physically bound water from crystallization in the temperatures below zero. The surfactants are dissolved in a mixture of low-boiling solvents of adequate partition coefficient between the phases of the system (Log $P_{O/W}$). Repeated application of the additive leads to total removal of water present in the tank. The problem of valve freezing is solved this way, and the corrosion threat is removed or at least minimized.

This product has been used by the largest local LPG producers for years. A water solubilizing agent, called Aquagasol, and an anti-corrosion additive, Corimsol, besides good opinions from the customers obtained a first grade award in the New Era Drive competition, in the category of innovative products.

Thanks to the application of Aquagasol additive, LPG contains water in concentration related to its production technology and operations in the distribution chain. The presence of water does not affect negatively the combustion process which is even improved.

4.1. Combustion of LPG with the addition of water

An interest in adding water to the combustion of hydrocarbon fuels has been observed since in 1913. Hoptinson introduced atomized water do the combustion chamber of a compression-ignition engine which resulted in the reduction of detonative combustion [14].

Water to the combustion process can be supplied in three ways: as an additional stream independent from the fuel, with the air, or in the form of fuel-water emulsion. It was demonstrated that, from the three mentioned methods, the use of a fuel-water emulsion offers biggest advantages. The use of such emulsion causes more efficient reduction of soot particles than the introduction of water in air. Moreover, Cornet and Nero [15] demonstrated that the use of an emulsion notably influences the increase of fuel efficiency and decrease of the temperature of emitted exhaust gases. During studies on the combustion of water-fuel emulsions also, the decrease of nitrogen oxides emissions has been observed [16, 17].

Addition of water as a separate stream, in the form of emulsion or in the form of vapors, has been profoundly studied [18, 19, 20, 21, 22]. Studies carried out in order to reduce harmful components of exhaust gases indicate that the addition of water to the combustion process is a promising measure allowing reaching these results. It was demonstrated experimentally that the addition of water to the combustion process gives significant effects for the combustion process of liquid fuels. Harrington [18] has proven that the addition of water to the combustion process can suppress detonative combustion and reduce the emission of nitrogen oxides as an effect of reduction of combustion temperature. He also observed that the concentration of carbon (II) oxide did not change and the concentration of hydrocarbons increased slightly in the exhaust gases emitted to the atmosphere. Özcan and Söylemez [23] carried out a research work aimed at the evaluation of addition of water to the combustion process of LPG in a traditional, spark-ignited engine. In these studies, water was introduced to the suction collector. The results demonstrated that the addition of water to the suction manifold causes the cooling of air-fuel mixture, slows the combustion speed reducing the top combustion temperature, which, in turn results in the reduction by 35% of maximum NOx emission, not changing CO and HC emissions. In Söylemez's and Özcan's work [24], the results of studies on the effect of water addition to the suction manifold on engine efficiency and the temperature of exhaust gases were presented. The results indicate that addition of water to suction manifold reduces the work of compression. Along with the increase of water-fuel weight ratio, the engine's torque, power, and thermal efficiency increased. Average increase of thermal efficiency for the water-fuel weight ratio of 0.5 was about 2.4% compared to the use of pure LPG for a tested range of an engine's rotational speeds. It was also found that along with the increase of the water-fuel ratio, the unitary fuel consumption and the temperature of exhaust gases decrease.

Addition of water to the combustion process of LPG causes the reduction of emission of harmful and toxic components of exhaust gases (NOx, hydrocarbons, CO2) and the reduction of fuel consumption.

An effect of water addition was studied in a spark-ignited engine fueled with liquefied gas on the temperature of exhaust gases and emission levels, for different fuel-air (F/A) ratios. The amount of formed nitrogen oxides (NOx) depends on the fuel/air ratio, the maximum temperature, and combustion velocity. The results indicate that the maximum, 35% reduction of NOx emission, is obtained by the introduction of water to the combustion of a lean mixture. The main cause of NOx reduction is the decrease of temperature and the reduction of combustion velocity caused by the addition of water.

For a lean mixture, the concentration of carbon (II) oxide changes slightly with balanced fuel/air ratio, and increases dramatically when the fuel-air mixture becomes rich, surpassing the stoichiometric ratio. Water in the combustion process improves complete combustion by promoting carbon (II) oxide oxidation. The combustion of CO is slow and results in the delay in fuel combustion. Water vapor in the zone of hydrocarbon combustion seems to exhibit catalytic action in this process. In this way, supplying water to the combustion chamber causes the acceleration of the combustion process [25].

A research has been conducted [26] which was aimed at the evaluation of water addition to the process of LPG combustion in a spark-ignited engine on the emission of particular components of exhaust gases, fuel consumption, and the level of fogging of exhaust gases. In this work, liquefied gas was used who's composition was determined chromatographically (Agilent 6890N) (Table 2).

Component	**% mol.**	**% wt.**
Ethane	0.34	0.20
Propane	45.93	39.26
Propene	0.00	0.00
n-butane	52.77	59.47
Isobutane	0.77	0.87
Isobutene	0.19	0.21

Table 2. Composition of LPG fuel

In this work, the emission of exhaust gases during the New European Driving Cycle (NEDC) was tested. This cycle is presently confirmed in the regulations and commonly used test in the measurements of exhaust gas emissions for passenger cars and light utility vehicles in Europe [27]. Obligatory for all newly produced cars in Europe, the NEDC test consists of two parts: urban driving cycle, representing fully driving in the city (UDC), 4 times 195 s, and extra-urban driving cycle, lasting for 400 s [28].

For this study, a passenger car Daewoo Lanos, produced in 2000, equipped with a 16 valve, 1498 cm^3 spark-ignited engine with an oxidation catalyst has been used.

The measurements of the emissions of exhaust gas components and fuel consumption were carried out in the Laboratory of Engines and Undercarriages, Automotive Industry Institute in Warsaw, equipped with a Schenck–Komeg EMDY chassis dynamometer with one roller of 48" diameter and MEXA 7200 exhaust gas analyzers produced by Horiba, including the following detectors:

- CO – AIA 7217 (NDIR method)
- CO_2 – AIA 7220 (NDIR method)
- O_2 – MPA 720
- HC – FIA 725 A (HFID method)
- NO_x – CLA 755A (chemiluminescence method)

as well as with a Opacimetr AVL 4390–G003 dynamometer.

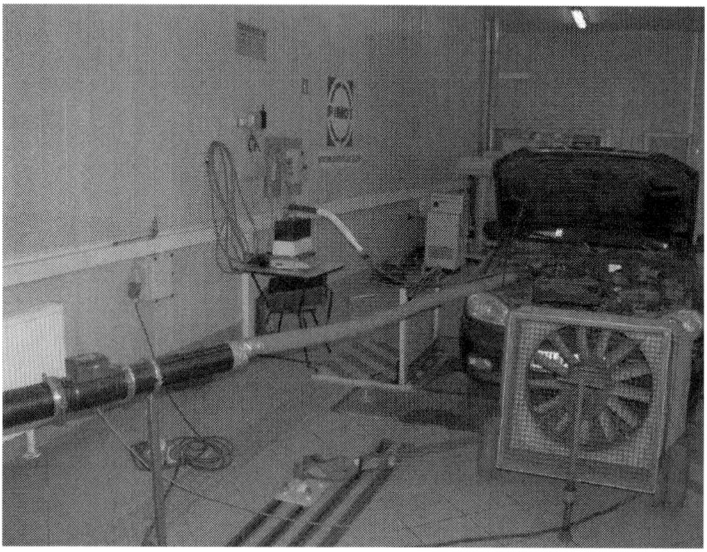

Figure 7. Photography of a test stand (visible chassis engine test stand, air supply, water vapor supply, and exhaust gas analyzer)

The amount of fuel and water introduced into the combustion chamber was measured by the change of mass of gas cylinder and ultrasonic generator of water vapor located after the air filter, before the throttle. Water was introduced in the form of ultrasonically generated cold vapors.

http://dx.doi.org/10.5772/61736

The studies encompassed the following types of supplying the fuel to the engine:

- LPG
- LPG with the addition of 2.22% m/m of water (LPG + water vapor (dry steam)).

Emission

The results of hydrocarbon emissions in exhaust gases for the two types of fuel supply to the engine, that is, pure LPG and LPG with the addition of water vapor, are presented in Table 3.

	UDC	EUDC	NEDC
LPG	3.43	0.24	1.42
LPG + water vapor	2.88	0.23	1.21

Table 3. The results of measurement of hydrocarbon emissions (g/km)

In the mixed cycle, NEDC, similarly as in the case of UDC and EUDC, the emission of hydrocarbons is reduced as an effect of addition of water vapor to the combustion process in the engine.

The results of carbon and nitrogen oxides as well as particulate matter emission measurements are presented in Tables 4, 5, 6, and 7.

	UDC	EUDC	NEDC
LPG	10.86	1.34	4.86
LPG + water vapor	7.46	0.63	3.15

Table 4. Emission of CO, g/km

The data presented in Table 4 indicate that the lowest level of CO emission is observed in the EUDC cycle. The largest CO emission is observed in the urban drive cycle, in which the emission of CO for LPG is 10.86 g/km and is reduced by the addition of water vapor to 7.46 g/km. The emission in the mixed cycle resembles the trend characteristic for the urban driving cycle.

	UDC	EUDC	NEDC
LPG	206.81	107.93	144.46
LPG + water vapor	191.82	104.08	136.41

Table 5. Emission of carbon dioxide, g/km

Emission of carbon dioxide reaches the highest level (206.81 g/km) in the case of urban driving on pure LPG and is reduced by the addition of water vapor to 191.82 g/km. This is the highest drop among the tests carried out of CO_2 emission.

	UDC	EUDC	NEDC
LPG	0.12	0.18	0.16
LPG + water vapor	0.11	0.17	0.15

Table 6. Emission of nitrogen oxides, g/km

From the comparison of data presented in Table 6, it can be concluded that the maximum NOx emission occurs in an extra-urban driving cycle, when the engine is supplied with pure LPG, reaching the value of 0.18 g/km. Addition of water vapor results in the reduction of NOx emission by a few percent in each studied cycle.

For each cycle, after the addition of water vapor to the mixture supplying the engine, a significant drop in the emission of particulate matter is observed, depending on the type of test. The biggest reduction of particulate matter emission was noted for the extra-urban driving cycle (from 0.0081 g/km to 0.0034 g/km) (Table 7).

	UDC	EUDC	NEDC
LPG	0.0071	0.0081	0.0072
LPG + water vapor	0.0039	0.0034	0.0036

Table 7. Emission of particulate matter, g/km

Fuel consumption

Table 8 presents fuel consumption, depending on the type of driving cycle

	UDC	EUDC	NEDC
LPG	15.21	7.21	9.62
LPG + water vapor	13.81	6.88	8.92

Table 8. Fuel consumption, dm^3/100 km

The analysis of this data allows the conclusion that addition of water vapor to the combustion process of LPG causes the reduction of fuel consumption. It is particularly visible in the urban driving cycle.

The analysis of the results of engine tests with the addition of water to the combustion in a spark-ignited engine fueled with autogas (LPG) confirms literature data, indicating that water

http://dx.doi.org/10.5772/61736

generally has good influence on the course of LPG combustion. Even a low concentration of water vapor in the fuel-air mixture causes significant reduction of toxic components in exhaust gases as well as reduction of fuel consumption. Particularly significant and important from the point of view of ecology is the reduction of particulate matter emission (Figure 8).

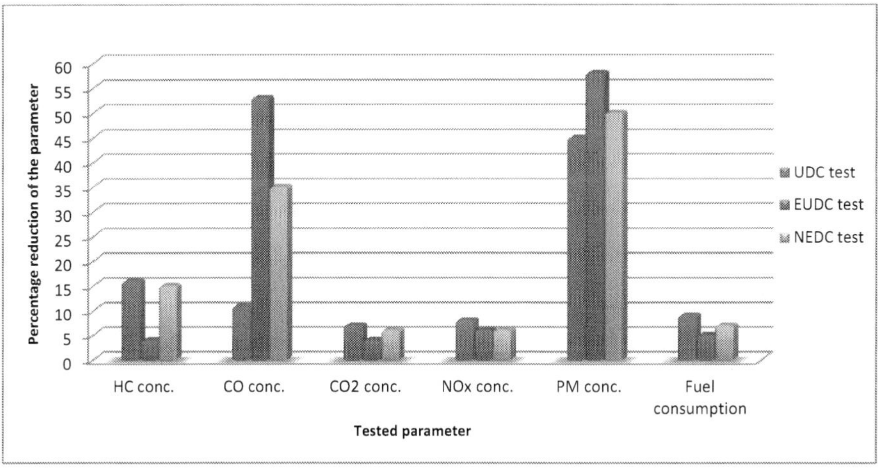

Figure 8. Percentage reduction of emission of exhaust gas components and fuel consumption depending on the type of test

The way of water addition to the LPG combustion process in a spark-ignited engine proposed in this work, turned out to be fully efficient and justifies further engine tests. Their aim will be the optimization and regulation of the amount of water added to the combustion of autogas as well as elaboration and construction of a water supply system.

5. Upgrading LPG as a motor fuel

A dominating way to use LPG as a motor fuel, at the present state of technology, is its controlled (sequential) dispensing into the suction manifold in gaseous state. This solution is rather complicated and not very rational. It requires liquid LPG to be evaporated by heating with cooling liquid, which requires starting of a cold engine on gasoline. Secondly, evaporated LPG cannot cool or cleanse the injector, which also leads to periodical dosing of gasoline to the engine fueled with autogas. Thirdly, such a form of LPG dispensing makes the use of additives more difficult, additives that could remove its disadvantages compared to gasoline.

A partial solution for the improvement of autogas quality, used in traditional fuel supply systems in spark-ignited engines, can be the use of a multifunctional additive [29]. The composition used in this invention significantly improves fuel lubricity and gives it adequate cleaning properties. The additive should generally be used in 50 ppm m/m concentration, which requires very high purity of the base fuel (oil residue). It performs very well in the storage and distribution of LPG, providing most of all longer lifetime of pumps (lubricity) and cleanliness of pipelines, tankers, and fittings (cleaning properties). Fuel improved with such an additive is ideal for modern injection systems of liquid LPG into suction collectors or directly into the combustion chamber [30].

The given example indicates the need for further improvement of LPG quality in order to meet or even exceed modern gasolines in all aspects, including operating parameters.

The users of LPG-fueled cars notice the differences in the quality of a fuel related to its calorific value. Because of this they most often blame the owners of fuel stations and attribute them with bad will or lack of care toward customers. They exhibit in this way the lack of understanding of an important feature of this fuel, which is the dependence of density on the temperature and chemical composition.

6. Modeling of LPG density as a function of composition and temperature

The problem originates from dual types of measurements used in the LPG distribution chain – based on mass and volume. Producers, importers, and wholesalers use measurements based on mass while fuel stations are based on volume. The solution of this problem could be mass counters in the LPG distributors at the filling stations. Modern, precise mass counters are presently adapted for high mass flows and are used most often in fuel bases and wholesalers; however, the producers currently make also equipment adequate to count smaller amounts of this fuel [31].

At the intermediate stages of distribution, the algorithms for the conversion of mass units into the volume of the fuel and vice versa are used. Both in the law and internal regulations, there are no detailed descriptions of correlations between these quantities, connected with density. Because of both practical and scientific reasons, this problem was the subject of cooperative work between the leading producer of LPG in Poland – Orlen Gaz Ltd., a part of the PKN Orlen capital group – and the Institute of Chemistry, Branch of Warsaw University of Technology in Plock.

In the first stage of this study, the producer picked out 10 batches of LPG of highly varying hydrocarbon composition (Table 9).

The search for a mathematical model for LPG density as a function of hydrocarbon composition and temperature leads to the problem of determination of the functional dependency in a 7-dimensional space:

$$\rho = f\left(x_1, x_2, x_3, x_4, x_5, x_6, T\right)$$

This dependency can be searched for using, for example, interpolation or approximation.

Batch No.	Propane	Propene	Isobutane	n-butane	Butenes	Isopentanes	Empirical density
[-]	[%m/m] x_1	[%m/m] x_2	[%m/m] x_3	[%m/m] x_4	[%m/m] x_5	[%m/m] x_6	kg/dm3 Q
1	38.8	9.2	25.8	21.6	3.9	0.5	0.5585
2	48,9	1,0	12,8	36,4	0,1	0,1	0.5590
3	47,4	2,6	18,5	29,4	1,2	0,2	0.5540
4	62,7	0,1	13,7	21,8	0,1	0,4	0.5455
5	48,0	4,3	11,7	26,4	8,9	0,2	0.5570
6	32,3	5,4	23,3	29,0	9,0	0,2	0.5640
7	33,0	3,6	20,5	28,6	12,9	0,4	0,5640
8	81,3	0,1	6,6	10,0	0,8	0,2	0,5320
9	83,8	0,5	4,9	6,9	3,2	0,1	0,5275
10	58,2	1,0	8,3	20,3	10,6	0,2	0,5510

Table 9. Hydrocarbon composition and density of LPG in 20°C (293,15K) [32]

After carrying out verifications of numerous mathematical models, a high susceptibility of numerical data (Table 9) to linear and exponential modeling has been noticed.

Assuming the following notifications:

T – temperature [K],

x_i – elementary composition of LPG fuel $\left[\frac{kg}{dm^3}\right]$,

w_i – determined weight coefficients [–],

b – determined shift coefficient [–].

Linear model assumes linear dependency of LPG density and chemical composition, which is described by the formula:

$$\rho = T \cdot x + b$$

where $x = w_1 \cdot x_1 + w_2 \cdot x_2 + w_3 \cdot x_3 + w_4 \cdot x_4 + w_5 \cdot x_5 + w_6 \cdot x_6$

or $\rho = T \cdot \sum_{i=1}^{6} w_i \cdot x_i + b$

The exponential model is based on an exponential function and assumes exponential dependency of LPG density on the temperature and chemical composition, which is described by the formula:

$$\rho = T \cdot e^{-x} + b$$

Having the value of real (empirically measured) density Q for the hydrocarbon composition of a batch numbered k, the determination of predicted density, ρ, requires the determination of weight coefficients w_1, w_2, w_3, w_4, w_5, w_6, using the least squares method.

Assuming $\varepsilon = \sqrt{(\rho - Q)^2}$ as an error of obtained approximation, the minimum of ε has to be found.

For the linear model, in order to determine the weight coefficients w_1, w_2, w_3, w_4, w_5, w_6 and shift coefficient b, the minimization of the formula $\varepsilon = \sqrt{\left(\left(T \cdot \sum_{i=1}^{6} w_i \cdot x_i + b\right) - Q\right)^2}$ has been carried out.

For the exponential model, the formula $\varepsilon = \sqrt{((T \cdot e^{-x} + b) - Q)^2}$ was minimized, where $x = \sum_{i=1}^{6} w_i \cdot x_i$.

Example of the results for the data presented in Table 9.

- For the linear model (for simplification b=0)

1,77E-05	1,81E-05	1,81E-05	2,14E-05	2,09E-05	8,56E-05
w_1	w_2	w_3	w_4	w_5	w_6

Table 10. Weight coefficients for linear model of LPG density and shift coefficient b=0

Projected density, for example, for the chemical composition number 4 is: $\rho = 293,15 \cdot (1,77 \cdot 62,7 + 1,81 \cdot 0,1 + 1,81 \cdot 13,7 + 2,14 \cdot 21,8 + 2,09 \cdot 0,1 + 8,56 \cdot 0,4) \cdot 10^{-5}$
$\rho = 0,54578$

- For the exponential model

0,164081	-0,042763	0,126409	0,070149	0,190154	0,092549	0,555551
w_1	w_2	w_3	w_4	w_5	w_6	b

Table 11. Weight coefficients for exponential model of LPG density of composition 4

Projected density, for example, for the chemical composition number 4 using the exponential model is:

$$x = 0{,}164081 \cdot 62{,}7 - 0{,}042763 \cdot 0{,}1 + 0{,}126409 \cdot 13{,}7 + 0{,}070149 \cdot 21{,}8 + 0{,}190154 \cdot 0{,}1 + \\ +0{,}092549 \cdot 0{,}4 = 13{,}60071131$$

$$\rho = 293{,}15 \cdot e^{-13{,}60071131} + 0{,}555551 = 0{,}555914466$$

Batch number	Q	Linear model		Exponential model	
ρ	Q - ρ	ρ	Q - ρ		
1	0,5585	0,5556	0,0029	0,5584	0,0001
2	0,5590	0,5589	0,0001	0,5571	0,0019
3	0,5540	0,5561	-0,0021	0,5569	-0,0029
4	0,5455	0,5458	-0,0003	0,5559	-0,0104
5	0,5570	0,5563	0,0007	0,5564	0,0006
6	0,5640	0,5636	0,0004	0,5578	0,0062
7	0,5640	0,5687	-0,0047	0,5568	0,0072
8	0,5320	0,5296	0,0024	0,5556	-0,0236
9	0,5275	0,5279	-0,0004	0,5556	-0,0281
10	0,5510	0,5486	0,0024	0,5558	-0,0048

Table 12. Summary of empirical and projected densities for the data presented in Table 10, along with approximation errors

For the needs of work [33] an application was elaborated.

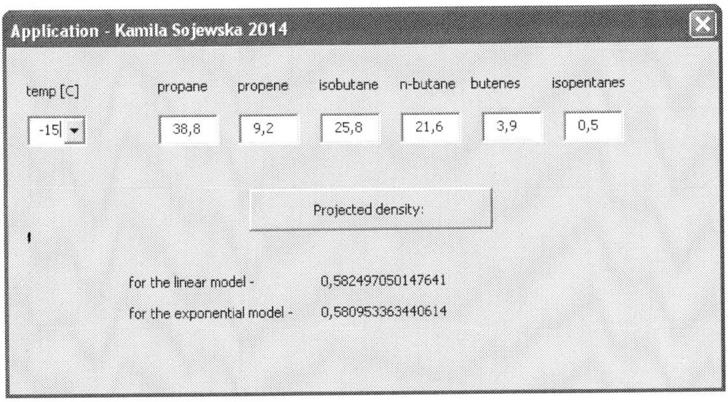

Figure 9. Application for the determination of LPG density basing on the chemical composition

Method of artificial neural networks

Using a thermodensimeter, the density of 10 LPG samples of different chemical compositions (Fig. 10) has been measured in the temperature range of –15ºC to 20ºC, every 1ºC [32]. In this way, a set of experimental data was prepared containing 360 different values of LPG density [g/cm³].

Temp [°C]	Batch No	propane	propene	isobutane	n-butane	butenes	isopentan	Empirical density
-15	1	38,8	9,2	25,8	21,6	3,9	0,5	0,5830
-14	1	38,8	9,2	25,8	21,6	3,9	0,5	0,5815
-13	1	38,8	9,2	25,8	21,6	3,9	0,5	0,5800
-12	1	38,8	9,2	25,8	21,6	3,9	0,5	0,5800
-11	1	38,8	9,2	25,8	21,6	3,9	0,5	0,5790
-10	1	38,8	9,2	25,8	21,6	3,9	0,5	0,5785
-9	1	38,8	9,2	25,8	21,6	3,9	0,5	0,5770
-8	1	38,8	9,2	25,8	21,6	3,9	0,5	0,5755
-7	1	38,8	9,2	25,8	21,6	3,9	0,5	0,5745
-6	1	38,8	9,2	25,8	21,6	3,9	0,5	0,5725
-5	1	38,8	9,2	25,8	21,6	3,9	0,5	0,5725
-4	1	38,8	9,2	25,8	21,6	3,9	0,5	0,5700
-3	1	38,8	9,2	25,8	21,6	3,9	0,5	0,5695
-2	1	38,8	9,2	25,8	21,6	3,9	0,5	0,5685
-1	1	38,8	9,2	25,8	21,6	3,9	0,5	0,5675
0	1	38,8	9,2	25,8	21,6	3,9	0,5	0,5655
1	1	38,8	9,2	25,8	21,6	3,9	0,5	0,5645
2	1	38,8	9,2	25,8	21,6	3,9	0,5	0,5635
3	1	38,8	9,2	25,8	21,6	3,9	0,5	0,5625

Figure 10. Example of data set uploaded to the Statistica software

The data set prepared in this way was uploaded to the Statistica software and processed using artificial neural networks. The variable number 9, density, is treated as a starting value, a dependent variable. The other columns, describing the hydrocarbon composition of LPG, are treated as independent variables. In such an arrangement, the density of LPG is expressed as a value dependent on the temperature (in Kelvin or Celsius) and six variables describing the composition of the fuel.

As an effect of using Statistica Automated Neural Networks (SANN) procedure, three multilayer networks of the Multi-Layer Perceptron (MLP) type were obtained.

MLP network of 7-7-1 layer architecture turned out to be an optimal choice. Artificial neural network taught in this way becomes a support tool contributing to the projection of LPG density depending on hydrocarbon composition at a given temperature.

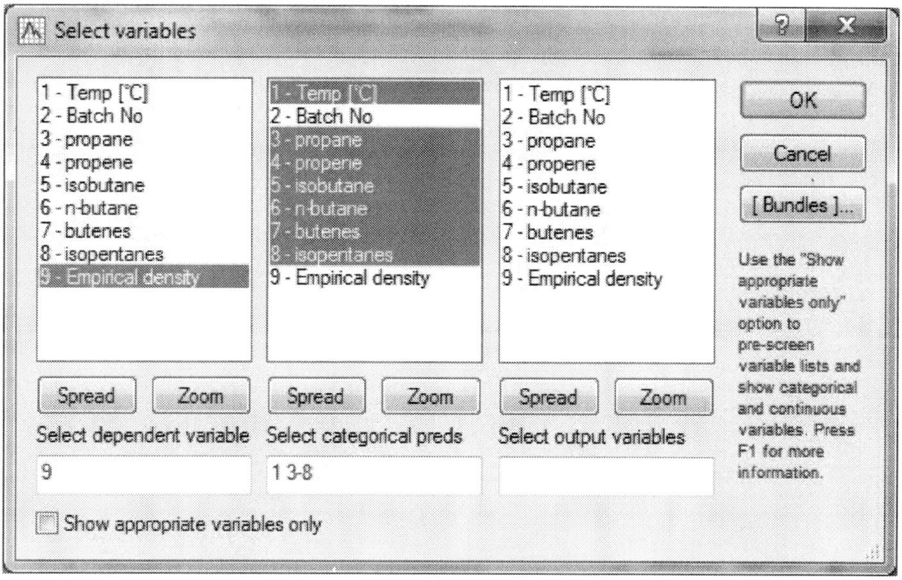

Figure 11. Interpretation of uploaded data

Empirical density Q	MLP 7-7-1 result ρ	Q - ρ
0,5830	0,5829	0,0001
0,5815	0,5818	-0,0003
0,5800	0,5807	-0,0007
0,5785	0,5775	0,0010
0,5770	0,5764	0,0006
0,5755	0,5753	0,0002
0,5745	0,5741	0,0004
0,5725	0,5719	0,0006
0,5790	0,5791	-0,0001
0,5780	0,5781	-0,0001
0,5765	0,5760	0,0005
0,5750	0,5750	0,0000
0,5740	0,5740	0,0000

Table 13. Comparison of densities, empirically determined and calculated by the neural network [34]

6.1. Analysis of a full spectrum of LPG hydrocarbon composition

An analysis seems justified of the influence of hydrocarbon composition on the LPG density in relation to the full spectrum of chemical compositions [35].

Batch number	Ethane	Propane	Propene	i-butane	n-butane	1-butene	i-butene	2-butene	i-pentane	Density in 15ºC [kg/m³]
1	0,2	99,1	0,6	0,1	0,1	0,1	0,1	0,1	0,1	0,507
2	0,2	35,7	1,6	23,7	36,3	0,6	0,1	0,7	0,5	0,549
3	0,3	79,9	0,8	8,1	9,8	0,3	0,1	0,4	0,2	0,519
4	1,3	49,1	0,2	19,9	29,1	0,1	0,1	0,1	0,3	0,536
5	0,2	59,3	0,2	14,9	25,1	0,1	0,1	0,1	0,3	0,532
6	1,2	31,9	0,2	29,6	35,9	0,8	0,8	0,1	0,1	0,548
7	0,2	75,9	0,5	9,7	11,3	0,8	0,1	0,8	0,2	0,521
8	1	49,3	0,1	18,5	30,4	0,1	0,1	0,1	0,1	0,537
9	0,2	62,2	0,5	14,2	22,2	0,1	0,1	0,2	0,2	0,531
10	0,1	49,7	0,2	18,2	31,6	0,1	0,1	0,1	0,1	0,539

Table 14. Examples of extended chemical compositions of LPG

Specification of butene content in liquefied gas leads to the search for functional dependency in a 10-dimensional space (including temperature):

$$\rho = f\left(x_1, x_2, x_3, x_4, x_5, x_6, x_7, x_8, x_9, T\right)$$

The use of numerical methods, interpolation or approximation, described earlier seems to be proper and justified. However, in the case of processing a huge set of experimental data, which guarantees adequate accuracy, this method generates too complex and expanded numerical models.

6.2. Classification of LPG basing on the propane content

LPG distributed in the winter season contains about 60% propane. It is necessary in order to provide adequate vapor pressure in low temperature. During summer, very high vapor pressure can cause evaporation already in fuel lines, and that is why in the summer season LPG contains about 40% propane.

The proposed method of LPG classification based on its composition categorizes liquefied fuels into one of the three groups, depending on the propane content [36]:

- Group I: above 60%
- Group II: from 60% to 40%
- Group III: below 40%

The proposed method of LPG classification, depending on the percentage share of propane, which, in turn, has influence on the vapor pressure, imposes automatically the division of analyzed empirical data into three groups. It results in turn with a smaller number of samples being subject at the same time to numerical calculations. It is directly connected with lower complexity and simplification of numerical analysis. This simplifies in turn the search for mathematical models expressing LPG density as a function of hydrocarbon composition and temperature.

Group	Ethane	Propane	Propene	i-butane	n-butane	1-butene	i-butene	2-trans butene	i-pentane	Density [kg/dm³]
I	0,2	99,1	0,6	0,1	0,1	0,1	0,1	0,1	0,1	0,507
	0,3	79,9	0,8	8,1	9,8	0,3	0,1	0,4	0,2	0,519
	0,2	75,9	0,5	9,7	11,3	0,8	0,1	0,8	0,2	0,521
	0,2	62,2	0,5	14,2	22,2	0,1	0,1	0,2	0,2	0,531
	0,2	71,3	0,4	15,9	0,1	0,1	0,1	0,1	0,1	0,524
	0,2	84,8	0,1	6	9	0,1	0,1	0,1	0,1	0,516
	0,6	60	9,2	21,2	6,6	0,4	0,6	0,7	0,2	0,525
	1,9	64	0,5	17,2	15,6	0,5	0,1	0,1	0,1	0,524
	0,4	61,7	8,8	12,5	14,7	0,8	0,1	0,5	0,2	0,526
	0,4	64	1,7	0,1	7,3	3,6	0,2	4,1	0,2	0,530
II	0,2	59,3	0,2	14,9	25,1	0,1	0,1	0,1	0,3	0,532
	1	49,3	0,1	18,5	30,4	0,1	0,1	0,1	0,1	0,537
	0,1	49,7	0,2	18,2	31,6	0,1	0,1	0,1	0,1	0,539
	1,1	41,3	0,3	25,2	27,9	3,4	0,6	0,1	0,1	0,542
	3,1	56,7	0,4	11,4	22,1	1,6	0,4	1,4	1,2	0,529
	3,5	56	0,2	13,8	25,3	0,1	0,1	0,1	0,7	0,527
	0,2	52,7	4,4	19,3	22,3	0,2	0,2	0,2	0,3	0,534
	1,5	58,1	0,2	16,8	21,7	0,1	0,1	0,1	0,7	0,530
	0,6	53,2	1,5	17,8	22,6	1,3	0,1	1,4	0,4	0,536
	0,8	49,7	11,4	24,2	10,2	0,8	0,7	1,1	0,2	0,531
III	0,2	35,7	1,6	23,7	36,3	0,6	0,1	0,7	0,5	0,549
	1	38,2	11,7	36,8	10,4	0,3	0,7	0,3	0,2	0,535
	0,4	32,4	10,8	42,9	10,4	0,4	1,3	0,7	0,1	0,542
	2,7	23,8	0,1	21,3	49,9	0,1	0,1	0,1	1,1	0,553
	2,6	23,3	0,1	21,2	50,1	0,2	0,1	0,1	1,1	0,554
	1	27,1	0,1	28,2	43,3	0,1	0,1	0,1	0,2	0,553
	1,1	36,8	1,4	35	10,7	12,4	2	0,1	0,1	0,545
	0,6	29,7	4,6	28,5	19,5	14,8	1,9	0,1	0,1	0,551
	0,8	33,5	1,7	10,9	36,7	0,1	0,1	9,2	0,2	0,555
	0,5	36,6	1,3	18,8	42,3	0,1	0,1	0,1	0,2	0,548

Table 15. Composition of the selected LPG samples, depending on the group, in the temperature of 15ºC

Investigating the linear model of the dependency of density on the hydrocarbon composition and minimizing the error of approximated density given by the formula:

$$\varepsilon = \sum_{k=1}^{10}\left[Q_k - T\cdot\sum_{i=1}^{9} w_i \cdot x_{ki} + b\right]^2$$

for each of three groups represented by 10 measurements, the values of weight coefficients, presented in Table 14, were obtained.

Weight coefficients										
w_1	w_2	w_3	w_4	w_5	w_6	w_7	w_8	w_9	b	
Group I	4,30 E-05	1,84 E-05	3,08 E-05	2,95 E-05	1,02 E-05	-3,45 E-04	-7,80 E-04	4,58 E-04	-8,69 E-04	1,00 E+00
Group II	2,26 E-05	1,66 E-05	1,62 E-05	1,66 E-05	1,46 E-05	1,10 E-05	6,66 E-06	2,39 E-05	4,09 E-05	1,00 E+00
Group III	2,83 E-05	1,74 E-05	1,70 E-05	1,45 E-05	1,45 E-05	1,31 E-05	2,59 E-05	2,28 E-05	1,95 E-05	1,00 E+00

Table 16. Weight coefficients for the linear model

Approximate density values, determined for each out of 10 empirical values, are presented in Table 15.

Group								
I			II			III		
Density Q	Density ρ	Q - ρ	Density Q	Density ρ	Q - ρ	Density Q	Density ρ	Q - ρ
0,507	0,5094	-0,0024	0,532	0,5323	-0,0003	0,549	0,5497	-0,0007
0,519	0,5170	0,0020	0,537	0,5383	-0,0013	0,535	0,5355	-0,0005
0,521	0,5211	-0,0001	0,539	0,5382	0,0008	0,542	0,5415	0,0005
0,531	0,5329	-0,0019	0,542	0,5420	0,0000	0,553	0,5522	0,0008
0,524	0,5243	-0,0003	0,529	0,5294	-0,0004	0,554	0,5547	-0,0007
0,516	0,5133	0,0027	0,527	0,5263	0,0007	0,553	0,5532	-0,0002
0,525	0,5250	0,0000	0,534	0,5338	0,0002	0,545	0,5449	0,0001
0,524	0,5241	-0,0001	0,530	0,5300	0,0000	0,551	0,5511	-0,0001
0,526	0,5259	0,0001	0,536	0,5355	0,0005	0,555	0,5550	0,0000
0,530	0,5301	-0,0001	0,531	0,5311	-0,0001	0,548	0,5472	0,0008

Table 17. Comparison of empirically determined and approximated density values

In Table 15, similarly as in previous cases, Q represents empirical density and ϱ approximated density. In order to summarize discussed methods of LPG density modeling on the basis of hydrocarbon composition at a given temperature, the differences between values of empirical density Q and approximated density ϱ are presented in Table 16.

	Linear model	Exponential model	MLP 7-7-1	Classification of LPG basing on propane content (group I, II, III)		
		$\lvert Q - \rho \rvert$		I	II	III
Min	0,00009	0,00009	0,00000	0,00000	0,00000	0,00000
Max	0,00469	0,02811	0,00100	0,00270	0,00130	0,00080
Median	0,00176	0,00550	0,00030	0,00020	0,00035	0,00050

Table 18. The comparison of the accuracy of discussed methods for predicting LPG density

7. Conclusions

In the past years, an increased interest has been observed in searching for such motor fuels, whose production and combustion results in lower emission of the main greenhouse gas, carbon dioxide. Because of environmental and economic considerations, the reduction of particulate matter, sulfur and nitrogen oxides emission, as well as the reduction of fuel consumption by vehicle engines are striven for.

From this point of view, a significant alternative for traditional motor fuels is fueling an engine with a liquefied mixture of propane and butane, called liquefied petroleum gas LPG or autogas. The advantages of liquefied hydrocarbon gases are particularly appreciated by the automotive industry.

On the background of increasing global LPG consumption, the achievements of Polish autogas industry in the area of production of automobile LPG systems, a number of bi-fuel vehicles, and logistical infrastructure were presented. The development of this industry is accompanied by research and implementation works, design and construction projects, as well as with promotion and marketing.

This fuel exhibits some disadvantages, the removal of which is a challenge to the autogas industry. The changes in the design of supply systems take place, and with regard to the fuel, quality improving additives are being tested, as well as addition of water to the LPG combustion process.

Successful solving of the problems indicated in this chapter will be a huge step forward to the improvement of competitiveness of LPG as a motor fuel.

In this chapter, several threads of innovative actions are presented, directed at further improvement of logistics efficiency, more complete utilization of advantages of LPG as a motor fuel, and indirectly the consolidation of a positive image of autogas as a fuel environmentally friendly.

Author details

Maciej Paczuski[1*], Marcin Marchwiany[2], Ryszard Puławski[1], Andrzej Pankowski[3], Kamil Kurpiel[4] and Marcin Przedlacki[1]

*Address all correspondence to: mpaczuski@pw.plock.pl

1 Institute of Chemistry, Plock Branch of Warsaw University of Technology, Poland

2 PKN ORLEN, Płock, Poland

3 Faculty of Civil Engineering, Mechanics and Petrochemistry, Plock Branch of Warsaw University of Technology, Poland

4 Control Service, Żurawica, Poland

References

[1] World LP Gas Association Annual Report 2013

[2] http://www.wlpga.org/about-lp-gas/production, accessed 25.06.2015

[3] Chłopek Z., "Paliwa alternatywne do silników spalinowych a emisja dwutlenku węgla kopalnego (Alternative fuels for internal combustion engines and the emission of fossil carbon dioxide)" Transport Samochodowy 2/2010. 37–47

[4] WELL-TO-WHEELS Report version 4.a : JEC WELL-TO-WHEELS ANALYSIS

[5] LPG Healthy Energy for Changing World, WLPGA, 2009

[6] Raport Roczny Polskiej Organizacji Gazu Płynnego (Annual report of Polish Liquefied Gas Organization), 2014

[7] http://www.ac.com.pl/pl/doc/295/o-firmie, accessed 25.06.2015

[8] www.gazeo.eu, accessed 25.06.2015

[9] Bukrejewski P., Skolniak M., Frydrych J. "Ocena istoty procesu pobierania próbek LPG na podstawie pierwszych w Polsce badań biegłości pobierania skroplonego gazu propan-butan (LPG) (Evaluation of the character of LPG samples collection, basing on pioneer in Poland research on the expertise of liquefied gas propane-butane (LPG) collection)", CHEMIK,2013,67,11,1122-1129

[10] Kurpiel K., "The Impact of the Structure of the Sampler for Sample Collection on the Representativeness of LPG Samples", Problemy Eksploatacji (Maintenance Problems), 2014,3,91-100

[11] Polish Patent PL 214183 (2013)

[12] http://www.petrolsc.com.pl/index.php?r=site/produkty&id=7#breadcrumb, accessed 25.06.2015

[13] http://www.petrolsc.com.pl/index.php?r=site/produkty&id=6#breadcrumb, accessed 25.06.2015

[14] Hopkinson P.B., I.M.E.Proceedings, 679 (1913)

[15] Cornet I., Nero E.W., Industrial and Engineering Chemistry, 1955, 2133

[16] Wang Ch.H., Chen J.-T., International Communications in Heat and Mass Transfer, 1996, 23 (6), 823-834

[17] Wilson J.P., "Effects of Water Injection and Increased Compression Ratio in a Gasoline Spark Ignition Engine". Wilson, Thesis, University of Idaho, 2011

[18] Harrington J.A., "Water Addition to Gasoline-Effect on Combustion, Emissions, Performance, and Knock", SAE Technical Paper, 1982

[19] Lanzafame R., "Water Injection Effects in a Single-Cylinder CFR Engine", SAE Technical Paper, 1999

[20] Abu-Zaid M., "Performance of Single Cylinder, Direct Injection Diesel Engine Using Water Fuel Emulsions", Energy Conversion and Management, 2004, 45, 697-705

[21] Kohketsu S., Mori K., Sakai K., Nakagawa H., "Reduction of Exhaust Emission with New Water Injection System in a Diesel Engine", JSAE Review, 1996, 17, 87-88

[22] Kegl B., Pehan S., "Reduction of Diesel Engine Emissions by Water Injection", SAE Technical Paper, 2001

[23] Syelmez S.M., Özcan H.,"Experimental Investigation of the Effects of Water Addition on the Exhaust Emissions of a Naturally Aspirated, Liquefied-Petroleum-Gas-Fueled Engine", Energy& Fuels, 2005, 19 (4), 1468-1472

[24] Syelmez S.M., Özcan H., "Water Injection Effects on the Performance of Four-Cylinder, LPG Fuelled SI Engine", Open Access Scientific Reports, 2013, 2 (1)

[25] Merkisz J., Piaseczny L., "Wpływ zasilania emulsją paliwowo-wodną na toksyczność i wskaźniki pracy okrętowego, średnioobrotowego silnika spalinowego (Influence of fuelling with a fuel-water emulsion on the toxicity and operation indicators of a navy, medium-rotational speed internal combustion engine)", Journal of KONES. Combustion Engines, 2001, 8 (3-4), 294-302

[26] Paczuski M., Marchwiany M., Bukrejewski P., Wójcik P., Biedrzycki J., "Sposób wprowadzenia wody do procesu spalania LPG, zwłaszcza za pomocą ultradźwię-

ków (The way of water introduction to the LPG combustion process, particularly with the use of ultrasounds)", Patent Application, 03/2015,26.03.2015

[27] Chłopek Z., "Ochrona środowiska naturalnego (Protection of natural environment)", WKŁ, Warszawa, 2004

[28] Romaniszyn M. K., "Badania porównawcze emisyjności samochodów zasilanych benzyną i gazem LPG (Comparative studies on the emissions from cars fuelled with gasoline and LPG) ", KONES 2000. 26 Międzynarodowa Konferencja Naukowa Silników Spalinowych, Nałęczów, 2000, s.290-293

[29] Pat. PL 397604 (2013)

[30] Pat. PL 399578 (2013)

[31] http://www.bronkhorst-cori-tech.com/en/products/coriolis_meters_controllers/mini_cori-flow_ex_d/, accessed 25.06.2015

[32] Pielat M., Engineering Thesis "Opracowanie modelu obliczeniowego zależności zmiany gęstości od temperatury paliwa LPG produkcji Orlen Gaz na podstawie wytypowanych składów mieszanin skroplonych węglowodorów naftowych (Elaboration of a mathematical model for the dependence between the density and the temperature of LPG fuel produced by Orlen Gaz basing on selected compositions of liquefied petroleum gases)", Instytut Chemii, Filia Politechniki Warszawskiej (Institute of Chemistry, Plock Branch of Warsaw University of Technology), Płock 2013

[33] Sojewska K., MSc Thesis "Modelowanie zależności gęstości LPG od temperatury i składu chemicznego (Modelling of the dependence of LPG density on its temperature and chemical composition)", Instytut Chemii, Filia Politechniki Warszawskiej (Institute of Chemistry, Plock Branch of Warsaw University of Technology), Płock 2014

[34] Bąbała W., MSc Thesis "Zastosowanie sztucznych sieci neuronowych do predykcji gęstości LPG w zależności od składu chemicznego paliwa i temperatury (Application of artificial neural networks for the prediction of LPG density as a function of its chemical composition and temperature)", Instytut Chemii Filia Politechniki Warszawskiej, Płock 2014

[35] Popławski A., MSc Thesis "Wpływ składu chemicznego na gęstość gazu płynnego LPG (The influence of chemical composition on the density of LPG)", Instytut Chemii, Filia Politechniki Warszawskiej (Institute of Chemistry, Plock Branch of Warsaw University of Technology), Płock 2015

[36] Kowalski M., MSc Thesis "Modelowanie gęstości LPG w zależności od składu chemicznego (Modeling of LPG density depending on its chemical composition)", Instytut Chemii, Filia Politechniki Warszawskiej (Institute of Chemistry, Plock Branch of Warsaw University of Technology), Płock 2015

Chapter 6

Microorganisms as Direct and Indirect Sources of Alternative Fuels

Anna Matuszewska

Additional information is available at the end of the chapter

http://dx.doi.org/10.5772/62397

Abstract

The industrialization and economic growth during the XXth century had been supported by fossil fuels, but it is clear that they are limited and they cannot sustain the growing energy needs. There is urgency in finding renewable and efficient fuels. The solar energy is obviously the solution in long-term but without suitable methods of storage, it is impossible to use it as a primary source of energy. One of the most important form of solar energy capturing is biomass itself, including the cell mass of microorganisms. The potential of microbes in alternative fuels and energy production is still unexploited. There are several possible routes for using a single-celled organisms to harvest energy.

The scope of this chapter include the review of possible technologies based on application of microorganisms in fuels and energy generation. The reader will find the reliable descriptions of currently available biotechnologies along with the ones that could become important in the future, like the new born technologies that are developed only on the laboratory stage.

The biotechnologies described in this chapter have been divided into two main groups, regarding to the role that the microbes play in the production process:

• the technologies in which the microorganisms serve as a direct source of biomass for fuel production from accumulated intracellular oil (e.g. microalgae and other oleaginous microorganisms biomass - bacillus, fungi and yeast),

• the technologies in which the abilities of microorganisms to excreting some valuable chemicals make them the indirect source of alternative fuels (e.g. methanogenic fermentation, ethanolic fermentation, fermentation of syngas etc.).

Additionally the methods of producing electrical power in microbial fuel cells (MFC) have been included, as a third group. In MFC, bacteria convert the energy from chemical compounds to electricity, that could be used as the final product or as driving force for other processes e.g. hydrogen production by microbiologically assisted water electrolysis.

The pros and cons of different presented scenarios, in which the microorganisms are playing the leading role as energy and fuel producers, have been mentioned. By giving the comprehensive basics for understanding the principles of wide range of processes, the author wants to introduce technologies that already exist and those which may be our future.

Keywords: Alternative fuels, microorganisms, microalgae, yeast, fungi, microbial fuel cells

1. Introduction

Since ancient times, humans used their surrounding nature as a source of food, medicine, energy and other valuable products. The discovery of coal and crude oil and their enormous energy potential has led people to focus their attention towards them. Oil and coal became the main raw materials for the production of energy and fuels for many years. They have been relatively cheap and easy to extract; moreover, the methods of their processing are well known. Rapid development in the industrial and automotive sectors caused crude oil resources depletion at a rapid rate. This resulted in the necessity of extracting crude oil from deeper layers, which caused increase in cost and influenced on the prices of petroleum products. Thus, the production of fuels from sources other than petroleum has become one of the main goals of humanity. The replacement of crude oil with other materials is very important for long-term energetic security and economic growth. Thus, once again the people turned their attention to nature and natural sources of energy such as water, sun or biomass. To meet the growing energy demand, people also began to use also microorganisms such as bacteria, microalgae, fungi and yeast. Today, these organisms are both raw material and producers of valuable substances used in many branches of industry, including production of fuels. Some microorganisms are able to convert waste biomass and biodegradable rubbish into desired products.

Due to the high diversity of microorganism species, their environmental requirements and eating habits, their ability to reproduce, yield of the desired compounds as well as their utilization and processing cannot be standardized as in the case of crude oil. There are several ways to obtain fuel products using microorganisms. Some of them are inexpensive, and today the production of biofuels is carried out on an industrial scale. Other technologies are in their infancy and require more time, more research, work of scientists and financial outlays to become ready for fuel production in the future.

In the literature, we can find some research involving microorganisms in production of biofuels, for example, production of biodiesel from microalgae or production of H_2 by bacteria. We can also find references that cover the revision of knowledge connected with particular methods of biofuel production or with processes taking place in the microbial bioelectrochemical systems (BESs). But there is a lack of literature, which includes the overall review of methods using microorganisms as a feedstock for the production of biofuels and as their producers.

http://dx.doi.org/10.5772/62397

2. Oleaginous microorganisms

Oleaginous microorganisms are species that are able to store oil in their cells, with the oil content excess of 20% biomass weight. This oil is commonly called *microbial oil*. It can be produced by some species of microalgae, bacteria, fungi and yeast. The main focus of research is on microalgae, fungi and yeast due to the yield of produced oil. Because the quantity of oil generated by bacteria is much lower, the interest in these organisms is not as high as for other microorganisms. The most advanced is work on technologies for producing biofuels from microalgae.

2.1. Microalgae

Microalgae are recognized as the oldest microorganisms present on Earth. They are present in all ecosystems and live under different environmental conditions. They are primitive unicellular form of plants. They do not have formed leaves, stems and roots. The cell structure of these species is very simple, which allows them to adapt to new environmental conditions relatively easy. The in-built chlorophyll in the cell of microalgae allows them to perform photosynthesis. But some species are heterotrophic and they require other sources of organic carbon and energy for growth. There are also mixotrophic microalgae, which, depending on the ambient conditions, can change their nutrition system from autotrophic to heterotrophic and vice versa.

Microalgae have a huge biodiversity; it is estimated that there are over 100 000 species of these organisms [1]. They differ among other species in cellular structure, life cycle and type of pigment. Two groups of microorganisms are classified as microalgae: prokaryotes and eukaryotes. The eukaryotes are divided into four main classes: the diatoms (*Bacillariophyceae*), the green algae (*Chlorophyceae*), the golden algae (*Chlorophyceae*) and the red algae (*Rhodophyceae*), whereas from prokaryotes to the microalgae group belong the cyanobacteria (*Cyanophyceae*) (Figure 1). Despite the large number of species of these algae, in practice, only about 15 species are used for large-scale production processes [2].

Under appropriate conditions such as temperature, the insolation level and nutrients, algae can grow abundantly. Typically, these organisms double their biomass for 24 hours, but their greatest growth phase (called exponential phase) is for 3.5 hours [3]. It is much higher than it is observed for agricultural crops or forestry. Because of such an intensive increase in biomass, microalgae need a lot of carbon, hydrogen and oxygen, which are taken from water and air. They also need minerals that are sources of nitrogen, potassium, calcium, magnesium, iron, sulphur, phosphorus, silicon and trace elements. All these elements are built in algae cells, so microalgae have become a source of these elements and of different valuable compounds such as pigments, lipids, sterols, fatty acids, starch, oils and others [4, 5].

Microalgae could be used to obtain different kinds of biofuels. The conversion of microalgae to energy can be realized in biochemical as well as thermochemical routes [6, 7] – Figure 2. We can obtain biodiesel from extracted microalgae oil, biomethane by anaerobic digestion of microalgae biomass, biohydrogen in dark fermentation stage or bioethanol by fermentation of

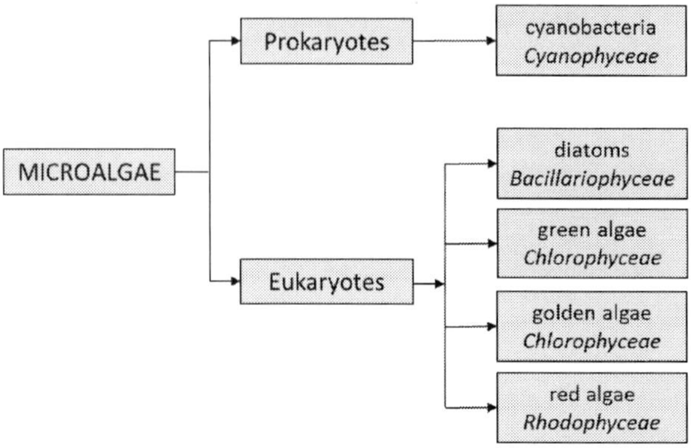

Figure 1. General division of microalgae.

ethanol. Algae biomass can be converted into biofuels not only biochemically but also thermochemically in processes such as pyrolysis, gasification and liquefaction. Another way to get energy is direct combustion of biomass.

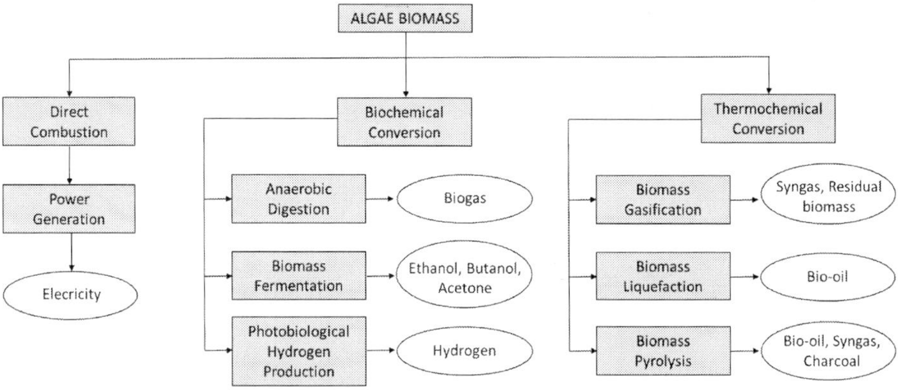

Figure 2. Potential routes for algae biomass conversion (adapted from [6]).

Regardless of the way of algae biomass conversion, the most important step is to provide the feedstock for a process. Microalgae are most often cultivated by special systems: open ponds and different kinds of photobioreactors [1]. The microalgae conversion process consists of several main steps: species selection, cultivation, harvesting, biomass concentration and algae pretreatment before conversion – Figure 3.

The very important stage in microalgae biomass production is the selection of the most appropriate species. These microorganisms can thrive in diverse environment such as fresh

http://dx.doi.org/10.5772/62397

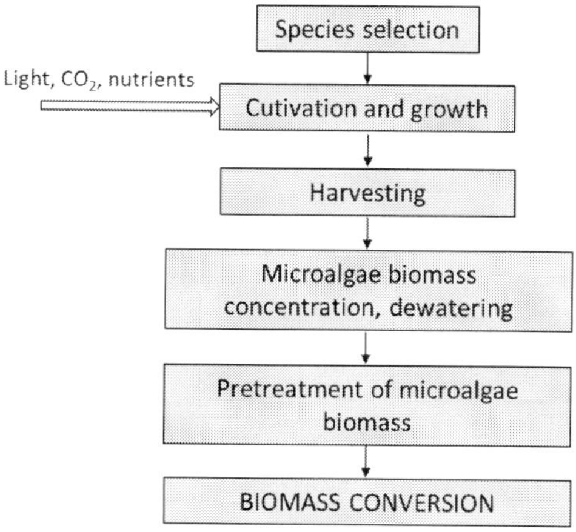

Figure 3. Scheme of microalgae feedstock providing.

water, brackish water or saline water. Currently, researches are being conducted on available species, but under artificial conditions these organisms lost their genuine properties due to their long-term cultivation. Such species are, for example, more sensitive to the environmental stress. In lab scale, the cells have more metabolites that improve the growth of algae [8]. On the other hand, in natural systems, microalgae can use other components that are not available under lab conditions. There is a lack of information about properties of species growing in isolate systems.

To select microalgae for any conversion process, the composition of cells and cell walls, productivity and the resistance to changes in temperature, pH, nutrition level, light or carbon supply should be taken into account. The kind of selected species and cultivation system (close or open ponds) decide the equipment, biomass yield, financial investment and so on. It should be remembered that microalgae production is much more expensive than oilseed crop production. These costs can be reduced, for example, using natural sunlight, CO_2 from industrial plants or nutrients (phosphates and nitrates) from sewage treatment plants or by changing the cultivation system.

2.1.1. Biodiesel

Generally, biodiesel is produced from oils extracted from oilseed crops or animals. These oils are esters of fatty acids and glycerine. Biodiesel is produced by transesterification, where the glycerine is replaced with other alcohol (most often methanol or ethanol) in the presence of a catalyst. The result of the transesterification process is a mixture of fatty acid methyl (or ethyl) esters. Since recently, microalgae are being considered as a potential source of oils for biodiesel production. They have the ability to accumulate lipids. Because lipids are basic raw material for biodiesel production, the oil content in cell should be taken into account for the microalgae selection process.

The oil content in cell depends on microalgae species and growth conditions [5, 8-10]. It can be as high as 80% dry weight. Examples of oil content of some algae cells are presented in Table 1. Microalgae need less land area for growing and have a higher yield of oil in comparison with oil crops. For example, oil yield from soybean is 446 L/ha, from palm oil 5950 L/ha, but from microalgae produced in photobioreactors, the oil yield is 58,700 L/ha for species with 30% oil by weight in biomass and 136,900 L/ha for species with 70% oil by weight in biomass [9].

Species	Oil content (% dry weight)
Botryococcus braunii	2575
Chlorella pyrenoidosa	2
Chlorella vulgaris	1440/56
Dunaliella salina	625
Dunaliella primolecta	23
Dunaliella tertiolecta	1771
Euglena gracilis	1420
Haematococcus pluvialis	25
Isochrysis galbana	740
Nannochloris sp.	2056
Nannochloropsis sp.	3168
Neochloris oleoabundans	2965
Phaeodactylum tricornutum	2030
Prymnesium parvum	2238
Tetraselmis maculata	3
Spirulina maxima	69
Scenedesmus obliquus	1222/3555
Schizochytrium sp.	5077
Tetraselmis suecica	923

Table 1. Oil content of some microalgae cells on the base of [1, 3, 9-11].

The difference between algae and other raw materials for biodiesel production is that the microalgae are microorganisms that generally live in water environments, and thus cultivation, harvesting and processing techniques are different. For mass diesel production, microalgae production is concentrated on one production unit where algae cells grow and then are separated from the growing medium and where the lipids are extracted from microalgae biomass. The biodiesel is produced in a way similar to existing processes and technologies used for other biodiesel feedstocks. Naturally, obtained oil can be used for production of other biofuels and then the transesterification reaction is replaced by other processes. For example, hydrotreating of vegetable oils is a quite modern way to produce very high-quality diesel fuels

[12, 13]. The other way could be a thermal decomposition or cleavage of the triglycerides and other organic compounds presented in the feedstock to get alkanes, alkenes and other chemicals [14, 15].

As mentioned earlier, feedstock providing is the important step before microalgae biomass conversion. Taking these stages into account, the process of biodiesel production from microalgae will be extended with additional operations such as oil extraction and biodiesel production. Figure 4 shows the main stages of biodiesel production process.

Each biodiesel used as a car fuel should meet the requirements of the International Biodiesel Standard for Vehicles (EN14214), also in terms of oxidative stability. The disadvantage of microalgae oils is that they contain much more polyunsaturated fatty acids with four or more double bonds than vegetable oils [16]. Such oils are susceptible to oxidation during storage, and therefore, their use for biodiesel production is limited. Many microalgae oils cannot be directly used as automotive fuels because of their composition, but the quantity of unsaturated bonds can be reduced easily by partial catalytic hydrogenation of double bonds [9]. The composition of microalgae oils depends on algae species; therefore, the proper selection of cultivated microorganisms is more important. The advantages of algae biodiesel over vegetable oil is that it is derived from plants that do not compete with foods.

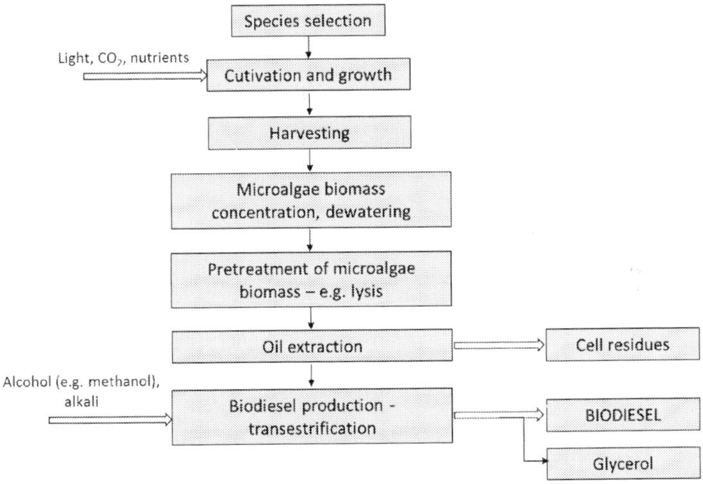

Figure 4. Scheme of microalgae biodiesel production.

2.1.2. *Biogas*

Apart from oils, microalgae in their cells also accumulate other components, for example, proteins, carbohydrates or starch, which are good nutriment for microorganisms producing biogas. Thus, these algae as well as maize silage or wastes from processing of fruit and vegetables can be a good feedstock for anaerobic digestion process. But comparing with other substrates in the literature, the studies on methane fermentation of microalgae are significantly less.

In recent years, two approaches to the production of biogas from microalgae have been investigated: (1) direct use of microalgae culture after concentration of the cells by filtration or centrifugation and (2) the use of cell residues, which remained after the extraction of oil or other components from microalgae.

Microalgae, due to the enormous growth, need very large amount of nitrogen and phosphorus to build their cells. In the case of nitrogen, annual demand is estimated between 8 and 16 tonnes/ha, which is 55 to 111 times greater than in the case of rapeseed [17]. It results in the fact that microalgae have a very big potential for the purification of water from the compounds containing N and P. The phosphorus and nitrogen accumulated in the cells remain in biomass after extraction of oil. Methane fermentation makes it possible to release these elements from the microalgae cells and then applying them as nutrients in the algae cultivation. It reduces costs of algae production, and the recovered biogas additionally improves the economy of the company.

In addition to the high content of N and P, microalgae contain many other minerals (Fe, Co, and Zn), which not only meet the nutritional requirements of anaerobic microorganisms, but also stimulate their growth. The content of above-mentioned compounds and minerals is differentiated and depends on the species of algae and conditions of their growth (in particular, the availability of nutrients). For example, deficiency of nitrogen in cultivation algae medium results in lower concentrations of proteins and higher concentrations of lipids. Thus, the yield of methane is also related to the type of microalgae used as a feedstock to the digestion process – Table 2 and Figure 5.

Species	**Growth conditions**	**Proteins [%]**	**Lipids [%]**	**Carbohydrates [%]**	**CH_4 [L CH_4 g VS^{-1}]**	
					Before lipid extraction	**After lipid extraction**
Chlorella vulgaris		29	18	51	0.64	0.56
Chlorella vulgaris	Low content of N	7	40	55	0.69	0.48
Chlorella emersonii		32	29	41	0.74	0.62
Chlorella emersonii	Low content of N	28	63	11	0.92	0.76
Chlorella protothecoides		38	11	52	0.65	0.60
Chlorella protothecoides	Low content of N	36	23	41	0.71	0.62

Table 2. The effect of low nitrogen content in the environment on the composition of three species of *Chlorella* and the theoretical methane potential before and after lipid extraction (adapted from [10]).

http://dx.doi.org/10.5772/62397

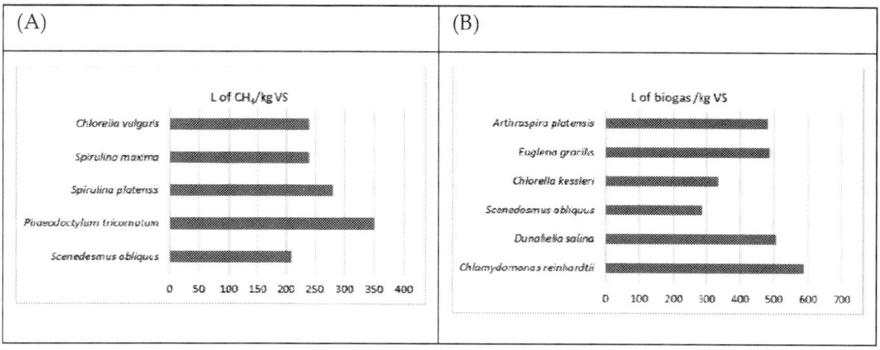

Figure 5. Examples of (A) methane and (B) biogas yields obtained during fermentation of some species of microalgae (adapted from [18]).

The yield of methane is determined not only by species of microalgae. It is important if to the fermentation process are used whole cells of or cells damaged, for example, by extraction of oils or other algae components. Methanogenic microorganisms have problem accessing the intracellular components because of the specific constitution of cell wall of algae (which can be different for various algae species). The destruction of the cell wall caused these components to become more available, resulting in an increase of biogas and methane yield. The extraction of lipids from microalgae cells before fermentation influences the composition of biogas. This treatment reduces lipids, which lowers methane yield (see Table 2) and increases the concentration of proteins. High content of proteins (as a result of extraction or due to kind of algae species) causes releasing of high amount of ammonium ions to fermentation medium. Higher concentration of those ions can inhibit fermentation process and, in extreme cases, may be toxic to the methanogenic organisms, especially at higher pH values. For example, fermentation of cyanobacteria, *Spirulina maxima,* which is rich in protein (contains up to 60% of proteins), resulted in release of large amounts of ammonia during hydrolysis (up to 7000 mg L^{-1}), resulting in withering anaerobic bacteria away [19, 20]. The advantage of biogas produced from microalgae is low hydrogen sulphide content in gaseous product.

When microalgae biomass is not subjected to any processes of destroying cell walls, the walls protect cells from the action of enzymes produced by hydrolysing bacteria. Thus, the biodegradation of the cells in the anaerobic process is weaker and biogas yield is smaller. Some species of microalgae can be much resistant to hydrolysis of their cell walls, for example, *Scenedesmus* and *Chlorella*. The cell walls of such algae have a multilayer structure and are mainly composed of organic compounds such as cellulose and hemicellulose, which are less biodegradable. Algal species devoid of the cell wall (e.g., *Dunaliella, Pavlova_cf*) or with a cell wall composed of glycoproteins (e.g., *Chlamydomonas, Euglena*) are more susceptible to microbiological degradation, resulting in higher yields of biogas and methane [21, 22]. The composition of the cell walls of microalgae is still poorly recognized and can considerably vary even within a given genus.

Theoretical yield of biogas is usually higher than real one because the bioavailability of feedstock determines biomass conversion during anaerobic digestion. Operations taken to increase the availability of the cellular content of microalgae for methanogenic bacteria allow for increasing the amount of gaseous product. As in the case, other types of biomass, degradation of the algae can be improved by the pretreatment. Therefore, there is a disintegration of the cell walls of microorganisms, which increases the availability of methanogenic bacteria to the contents of the cells and thereby affects the biogas productivity. The process of disintegration of microalgae can be carried out using different methods: thermal, mechanical, chemical or biological. The efficiency of used method depends on the parameters of the process (e.g., temperature, exposure time and dose of chemicals), as well as the characteristics of microalgae (the strength and structure of the cell wall and macromolecular composition of algae cells).

An important parameter during methane fermentation is the carbon (C) to nitrogen (N) ratio in biomass feedstock, which should range from 20/1 to 30/1. When this parameter is not properly balanced and the C/N is less than 20, the significant amounts of nitrogen (volatile fatty acids) are released, leading to their accumulation in the fermenter. Such a situation causes the inhibition of the methane production because of acidification of process environment. C/N ratio can be controlled by choosing the kind of substrates and their proportions in the mixture. Microalgae have low C/N value (below 10) [23], and it would be well to introduce an additional substrate rich in carbon (co-digestion) such as maize silage or sludge from sewage treatment plants. In such cases, not only the increase in methane yield was observed [24, 25], but there has been reported cases where the real yield was higher than the theoretical (synergy effect) [26, 27]. Co-digestion may also lead to the dilution of some compounds present in the microalgae biomass (e.g., sodium compounds) that have an inhibitory or toxic effect on the anaerobic bacteria.

2.1.3. Butanol and ethanol

Fermentation is a common process used commercially on a large scale to produce ethanol from crops containing sugar and starch. Microalgae accumulate polysaccharides in their cells. They are rich in various carbohydrates such as cellulose, starch, mannitol, agar and laminarin [28]. Some species contain a large amount (even over 50% of the dry weight) of starch and glycogen, which are essential for ethanol production. Such species include *Chlorella, Chlamydomonas, Dunaliella, Scenedesmus* and *Spirulina* [29, 30]. The examples of microalgae and the starch content in the cells are presented in Table 3. This causes these organisms can be used as a good raw material for alcohol production.

Ethanolic fermentation is performed mainly by yeast. Fermentation process for microalgae is similar to that for other plants. The biomass is ground down, and the polysaccharides are converted to monosaccharides. Then, the yeast breaks down the sugars and converts them to ethanol. Acetone–butanol fermentation is an anaerobic process of enzymatic degradation of saccharides to butanol, acetone, carbon dioxide and hydrogen. It is carried out by certain species of bacteria of the genus *Clostridium* (especially *C. Butylicum* and *C. acetobutylicum*).

Generally, the whole process of alcohol production consists of pretreatment of the biomass, saccharification, fermentation and product recovery.

Species	% Starch (g/dry weight)	Reference
Chlorella vulgaris	37	[29]
Chlorella vulgaris	17	[31]
Chlamydomonas reinhardtii	17	[31]
Chlorococcum humicola	11	[32]
Nostoc muscorum	34	[30]
Oscillatoria obscura	13	[30]
Scenedesmus obliquus	23	[30]
Spirulina fusiformis	37-56	[30]

Table 3. Starch content in the microalgae cells.

Quantity of produced alcohols depends on the production process parameters and characteristic of used algae biomass. For instance, the microalgae species such as *Chlorella* and *Chlorococcum* are better raw materials for ethanol production than *Chlamydomonas* [28]. As in the case of biogas production, the yield of alcohol is also dependent on the microalgae biomass pretreatment. During these operations, the fermentable sugars are released from algae cells and become available for the fermentation process. The biomass, as in the case of methane fermentation, can be pretreated in physical, biological and chemical ways. Harun and Danquah [32] studied an acid exposure as a pretreatment method to release the carbohydrates from cells of *Chlorococcum* species. The highest ethanol yield was obtained when the microalgae (in concentration 15 g L^{-1}) were treated at 140°C with 1% (v/v) of sulphuric acid for 25 minutes. They found that during this pretreatment method the most important parameter that influences bioethanol production from microalgae is temperature. The other significant parameters are the acid concentration and the amount of microalgae loading.

Ethanol from microalgae can be produced in three ways: (1) from algae cell components (starch and saccharides) after their extraction or from cell wall components (cellulose) after enzymatic hydrolysis of walls, (2) some species of microalgae product ethanol during dark fermentation and (3) via genetic modification of some microalgae to direct production of ethanol [30].

In the first way, the starch can be extracted from the cells using mechanical (ultrasonic and steam explosion) or biological methods (dissolution of cell walls by enzymes). The starch after separation is used for fermentation process using the technology similar to other feedstock rich in starch.

Alcohol can be also produced from the cell residues after oil extraction [33]. Harun et al. investigated the possibility of using the cells of *Chlorococcum* species after oil extraction to ethanol production. In their experiment, the ethanol yield was about 3.8 g L^{-1} from 10 g L^{-1} of the substrate.

Some microalgae species are rich in starch and their cell walls contain cellulose. In case of such algae, the cellulase-producing microorganisms are used to convert this carbohydrate into simple sugar and then the all biomass can be fermented to ethanol. Depolymerization of cell walls (chemical and biological) increases the amount of monosugars and thus the yield of ethanol. Some microalgae contain other sugars (e.g., mannitol and laminarin). It must be remembered that not all anaerobic bacteria are able to ferment all kind of sugars (e.g., mannitol). Therefore, it is very important to know the algae cell composition and find appropriate microorganisms for fermentation process [30].

The stage that influences ethanol yield is a kind of a fermentation process. Harun et al. [34] investigated three of them: (1) separate hydrolysis and fermentation (SHF), (2) separate hydrolysis and co-fermentation (SHCF) and (3) simultaneous saccharification and fermentation (SSF) with different pretreatment of *Chlorococcum* sp. biomass (acid and enzymatic hydrolysis). They stated that acid SHF was the most effective method of ethanol production, and the fermentation process conducted in continuous way is more efficient than batch fermentation.

In lack of light and in presence of oxygen, microalgae convert starch or glycogen by oxidizing them to carbon dioxide. But in dark, under anaerobic conditions, the oxidation is incomplete and different products (such as hydrogen, ethanol, formic acid and acetic acid) are produced. The proportion of particular compounds depends inter alia on the species of microalgae. Microalgae whose cells contain polysaccharides composed of glucose (e.g., *Chlamydomonas, Chlorella, Microcystis, Oscillatoria,* and *Spirulina*) are able to produce ethanol in the dark under oxygen-free conditions easily. The yield of alcohol production can be enhanced by appropriate pH and temperature range [35, 36]. Ueno et al. [36] obtained the maximum productivity of ethanol (450 μmol g^{-1} dry weight) from *Chlorococcum littorale* at 30°C. Beside the ethanol, the fermentation products were acetate, hydrogen and carbon dioxide. It was stated that ethanol productivity can be improved by adding methyl viologen, additionally resulting in decreased production of hydrogen (more electrons were involved in ethanol formation in the presence of methyl viologen, which is used as an electron acceptor and transfer catalyst in redox reactions). Hirano et al. [35] stated that intracellular ethanol production is simpler and less energy intensive in comparison with the conventional ethanol-fermentation process.

The processes of ethanol production by biomass pretreatment, extraction, fermentation and so on involve costs. It would be interesting if the microalgae can produce alcohol directly. Currently, intensive researches are conducted with the aim of increasing the accumulation of compounds (lipids, starch, alcohol, etc.) in photosynthetic organisms using genetic engineering. The glucose and other metabolites of algae are produced in Calvin cycle of photosynthesis. Using genetic engineering, the ethanol-producing genes from the ethanologenic bacterium *Zymomonas mobilis* were introduced to cyanobacterium *Synechococcus* sp. New species were able to produce ethanol in presence of light. The ethanol produced by the transformed cyanobacterium diffused from the cells into the culture medium [37]. This way of ethanol production is still in the research; there are more questions than answers.

2.1.4. Hydrogen

Microalgae have capacity for producing hydrogen by photobiological reaction. The hydrogen is produced by direct or indirect photolysis of water [38]:

Direct photolysis: $2\ H_2O \rightarrow light \rightarrow 2\ H_2 + O_2$

Indirect photolysis:

1. $12\ H_2O + 6\ CO_2 \rightarrow light \rightarrow C_6H_{12}O_6 + 6\ O_2$

2. $C_6H_{12}O_6 + 12\ H_2O \rightarrow 12\ H_2 + 6\ CO_2$.

During photosynthesis, water molecules are converted by microalgae into hydrogen ions (H^+) and oxygen. Then the hydrogen ions are converted into hydrogen with the use of hydrogenase enzymes. The presence of oxygen results in rapid inhibition of the hydrogenase enzymes and the hydrogen production process is impeded. Therefore, cultivation of microalgae for H_2 production must be realized under anaerobic condition [29].

Photosynthetic production of hydrogen can be carried out with the use of two-stage method. In this process the photosynthetic generation of O_2 and production of H_2 are separated. In the first stage, the algae grow photosynthetically under normal conditions. During the second stage, the access to the sulphur is limited and microalgae are exposed to anaerobic conditions. Under S deprivation conditions, microalgae are fundamentally altering photosynthesis and cellular metabolism to survive. They consume internal starch and protein and produce hydrogen. This production process is limited with time, the yield of hydrogen decreases after 60 hours of production. In this method, the theoretical maximum yield of H_2 production by green algae is 20g H_2 m^{-2} d^{-1} [39]. The use of this method for hydrogen production does not generate any undesirable, toxic or environmentally harmful by-products.

Another method for hydrogen production is a continuous mode. In this mode, electrons and protons that are released during photosynthetic H_2O oxidation are directly recombined by the hydrogenase to produce hydrogen. Theoretically, such a process is for 33% more efficient than two-phase method because in the two-phase process, electrons and protons released from water are storage (e.g., as starch) before being use to H_2 generation [40].

Technical and physiological parameters of microalgae cultivation influence hydrogen production efficiency. As described by Kruse and Hankamer [40], the important parameters are light intensity, chlorophyll concentration, culture mixing, pH and the interplay between these parameters.

2.1.5. Thermochemical conversion

Gasification, pyrolysis and liquefaction are basic processes of thermochemical conversion of microalgae biomass. Before these processes the valuable substances contained in the cells of algae are very often extracted. The principles of thermochemical conversion methods of microalgae are similar to these for other types of biomass.

Gasification is a process of partial oxidation of biomass with air, oxygen and/or steam at high temperatures, usually in the range 800-1000°C. The biomass is converted into gaseous product (syngas), which is the mixture of hydrogen, carbon oxide, carbon dioxide, methane and nitrogen. Nitrogen content can be reduced after mineralization to ammonia. Syngas has low calorific value (usually 4-6 MJ m^{-3}) and can be used as a fuel for gas turbines or engines [41]. The syngas is a crucial intermediate resource for production of other compounds such as methanol, synthetic hydrocarbons, ethanol and others via the Fischer–Tropsch (FT) process. It can also be microbiologically fermented where the main products are ethanol, butanol, butyric acid, acetic acid and methane. Due to substantial water content in microalgae biomass, before gasification, the feedstock can be partially dried. There are only a few works reported in literature concerning the gasification of microalgae. Hirano et al. [42] partially oxidized *Spirulina* at temperatures of 850°C, 950°C and 1000°C and determined the composition of obtained gas in order to evaluate the theoretical yield of methanol. They stated that the gas composition depends on the process temperature. The highest theoretical yield of 0.64 g methanol from 1 g of the biomass was obtained by them for the gasification conducted at 1000°C. Additionally, they estimated energy balance (ratio of the energy of methanol produced to the total required energy), which was slightly disadvantageous. The greater part of energy is used for algae cultivation, thus the balance can be significantly improved by more efficient production of microalgal biomass. López-González et al. [43] investigated gasification process of chars obtained from the pyrolysis process of three microalgae *Scenedesmus almeriensis, Nannochloropsis gaditana* and *Chlorella vulgaris* with the use of thermogravimetric–mass spectrometric analysis. They stated that the indigenous mineral matter present in microalgae samples catalytically influences the gasification process. The metals in microalgae samples influenced the samples' reactivity as well as the production of gases. The highest gas yields were obtained for *Scenedesmus* sample, which was characterized by high potassium content.

The other form of gasification of microalgae is catalytic supercritical water gasification SCWG – a kind of steam reforming. In this process, the thermal conversion of algae could be conducted for biomass with high moisture content (50-90%). SCWG is realized in lower temperatures (250-360°C), pressure about 20 MPa and in the presence of catalyst. The system is operated as a liquid-phase and the main product is a mixture of methane (50-60%) and carbon dioxide [44]. The problem in this method is presence of biomass trace components, which can react with the catalyst and significantly reduce its activity.

Catalytic supercritical water gasification conversion is characterized by a high chemical energy conversion efficiencies (up to 70-77%) and short time of reaction (order of minutes) so high rates of biomass conversion are possible on a much smaller area in comparison with anaerobic digestion. The SCWG process enables recovery of some nutrients from the microalgae biomass because the fluid shows low solubility for salts [45]. Cherad et al [46] gasified macroalgae *Saccharina latissima* in a batch reactor at 500°C and 36 MPa and studied the influence of biochemical content and ash on syngas composition. Such a process can also be used for microalgae. They stated that the presence of Ru/Al_2O_3 catalyst caused an increase in the yields of hydrogen (30%) and C1-C4 gases, and the gasification efficiency in comparison with a process without this catalyst. Their results also indicated that the process water recovered from gasification of microalgae can be used as nutrients during microalgae cultivation.

Another thermochemical process of microalgae biomass is its liquefaction (HTL process), which is similar to SCWG. It is direct hydrothermal liquefaction in sub-critical water conditions and can be employed to convert wet algal biomass into liquid fuel. The process is realized at low temperature (300–350°C), high pressure (5–20 MPa) and in the presence of catalyst and hydrogen. Sub-critical conditions enable decomposition of biomass materials to shorter and smaller molecular materials (bio-oil) with a higher energy density [29]. Hydrothermal liquefaction is considered for being the most promising technique for conversion of wet algal biomass. It has been shown that the yield of bio-oil from HTL of whole biomass exceeds the lipid content in the raw material [47]. Disadvantages of hydrothermal liquefaction are the complexity and very high cost of the apparatus. Another way to produce bio-oil is pyrolysis - heating the biomass in the absence of air at 500°C, without or in the presence of a catalyst for very short time. The main products are bio-oil, charcoal and gas. The ratio of particular products depends on hot vapour residence time and temperature. The shorter the time, the higher the yield of the liquid product. The high biomass-to-liquid conversion (yield of bio-oil about 75%) can be achieved during flash pyrolysis (hot vapour residence time about 1 second, temperature 500°C) [29]. Since algae contain a lot of moisture content, a biomass must be initially dried [7], which significantly increases the cost of bio-oil production. The HTL process enables to avoid this cost.

The composition of bio-oil depends on a kind of algae biomass, method of its conversion and parameters of process. Algae-derived bio-oil contains large amounts of heteroatoms, such as N, S and O. It contains long chain fatty acids, resulting in high viscosity, so such products cannot be used directly as a fuel [48]. But it can be converted together with crude oil or independently to other valuable products. The microalgae bio-oil could contain some metals such as Fe, Mg, Ni and Zn, which are present in original algae cells. These metals can cause some difficulties for the upgrading process. Galafassi et al. [49] stated that it is possible to remove oxygen and metals from crude algae oils produced by HTL only by thermal means without the use of catalysts or hydrogen. Thermal treatment can reduce amount of acids in the bio-oil, decrease its viscosity and make it more volatile. The higher temperature of thermal treatment reduces the amount of trace metals present in bio-oil more effectively.

2.2. Others oleaginous microorganisms

Microbial oil (called also single-cell oil SCO) can be obtained not only from microalgae. Some microorganisms such as yeast and fungi (especially moulds) or bacteria can also be a source of it. However, the use of these organisms as a source of lipids for biofuels production still remains in the sphere of research. Oleaginous yeasts and moulds produce polyunsaturated fatty acid triacylglycerol, which is similar to vegetable oils. Produced triglycerides are rich in polyunsaturated fatty acid such as oleic (C18:1), linoleic (C18:2), palmitic (C16:0) and palmitoleic (C16:1). Because the fatty acid profile of microbial oils is similar to that of plant oils, oleaginous yeast and fungi can be a favourable feedstock for the biodiesel industry, but currently, the production of microbial oil from these organisms is expensive. Similar to microalgae, yeasts accumulate reserve lipids as storage metabolites, especially they suffer a deficiency of nutrients, usually involving nitrogen, but with access to the carbon-containing components [49]. Thus, the lipid production depends on the carbon-to-nitrogen ratio (C/N). Nitrogen is used for production of nucleic acids and proteins, whereas carbon is necessary for energetic processes and synthesis of proteins, carbohydrates, nucleic acids and lipids. During nitrogen shortage, the growth of microorganisms slows down and production of nucleic acids, proteins and carbon is consumed for storage lipid synthesis [50].

In contrast to algae, yeast and fungi are not able to produce carbon compounds from CO_2, because they cannot carry out photosynthesis. They must obtain a carbon source (e.g., fat) from a medium in which they live. Oleaginous yeast or moulds growing in media, which contain fats as a carbon source, accumulate lipids in cells during primary anabolic growth. This accumulation process is not limited by presence of some nutrients in the cultivation environment. Although storage lipids are produced from glucose or other similar components, during second anabolic activity, the lack of some nutrients in the growth medium is a condition of accumulation process [51, 52]. In case of oleaginous microorganisms growing on fats, when the deficiency of fatty acids in growing medium have place, the organisms start to consume their own storage lipids for theirs metabolic requirements and growth.

The lipid content in oleaginous yeast and fungi depends inter alia on their species (Table 4) and growth conditions such as a type of carbon source, pH and temperature. Due to various accumulation of lipids, there are not many species of fungi and yeast, which can be applied as a feedstock for biodiesel production. For example, among over 600 species of yeasts, less than 30 are able to accumulate more than 25% of their biomass weight as oil [53]. Therefore, before using any yeast or fungi for biofuel production, the appropriate selection and characterization of oleaginous strains should be performed.

Fungi and yeast have several advantages over conventional plant and microalgae. Their cultivation is easy; they can be grown in bioreactors. Yeast and fungi have short life cycles, characterized by the rapid growth rates, which are unaffected by space, light or climatic variations. Their cultivation can be easily scaled up. They have been found to be robust microorganism that can grow on various substrates such as lignocellulosic biomass and agro-industrial residues, for example, glycerol, fats, whey and molasses [50, 51, 54]. Some species can also grow in sewage sludge, waste water or in salt water.

Extraction of lipids from fungi and yeast can be carried out with the methods that are used during oil extraction from microalgae, for example, cold extraction with solvents. Obtained oil can be used as a feedstock for biodiesel production by its transesterification. The amount of obtained oil can be increased by biomass pretreatment, but there is a lack of information about it. Tsigie et al. [57] treated *Yarrowia lipolytica* Po 1g yeast with subcritical water (SCW) method. It is an environment-friendly technique for increasing the amount of extractable lipids in microorganisms. Treatment is performed in water (in the liquid state) at temperatures ranging between 100 and 374°C under high pressure. This method enabled to increase the amount of extractable lipid from 51.53% to 84.75%. Results suggest that solvent extraction alone is not an effective method for the complete recovery of lipids from *Yarrowia lipolytica* Po 1g.

Species	Oil content (% dry weight)
Yeast	
Candida curvata	58
Cryptococcus albidus	65
Lipomyces starkeyi	64
Rhodotorula glutinis	72
Rhizopus arrhizus	57
Trichosporon pullulans	65
Yarrowia lipolytica	36
Fungi	
Aspergillus oryzae	57
Humicola lanuginosa	75
Mortierella isabellina	86
Mortierella alpina	40
Mortierella vinacea	66
Rhizopus stolonifer LGAM (9)1	28

Table 4. Oil content in the oleaginous microorganisms cells [53, 55, 56].

Dai et al. [58] showed that biodiesel obtained by transesterification of oil from *Rhodotorula glutinis* yeast possessed similar composition to that from vegetable oil. Other promising species for the microbial oil production is *Yarrowia lipolytica,* which is used in the industry for the production of citric acid and protein [52]. However, in lipids production from oleaginous microorganisms is a lot of unknown, for example, connected with metabolism of these organisms (e.g., functional and structural properties of enzymes). Not enough is yet known about the effects of nutrients in culture medium on the quality of oil produced in cells. Some substrates, for example, molasses, raw glycerol lead to accumulation higher quantity of saturated fatty acids in comparison with rapeseed oil [59]. On the one hand, this may result in the improvement of the cetane number and oxidation stability; on the other hand, it adversely affects the low-temperature properties and viscosity. High cost of cultivation, especially substrate cost, causes that commercialization of fuel production from fungi and yeast will not quickly occur. Genetic engineering, by improvement in lipids accumulation in oleaginous microorganisms or by generation strains which are able to produce specific fatty acid compositions, can improve economics of microbial oil production.

Similar to fungi, some species of bacteria can also accumulate lipids in the form of triacylglycerol under some special environment, but this oil is different from other microbial oil. Generally, the oil produced by most bacteria is complex; only a few species produce oil that can be a raw material for fuel production, and some examples are presented in Table 5.

Bacterium	Oil content (% dry weight)	Reference
Arthrobacter sp.	>40	[53]
Acinetobacter calcoaceticus	27-38	[53]
Bacillus alcalophilus	18-24	[53]
Gordonia sp.	80	[60]
Rhodococcus opacus	24-25	[53]
Rhodococcus opacus PD630	72	[60]

Table 5. Oil content in cells of some bacteria species.

As stated by Gouda et al. [60], *Gordonia* sp. and *R. opacus* PD630 are able to accumulate oils under special conditions with maximum oil content over 70%, but the biomass is only 1.88 g L^{-1}. The advantage of bacteria in comparison with microalgae is their higher growth rate and easier culture method [53].

The wealth of knowledge on the genetics and metabolic pathway of bacteria makes them ideal candidates to research with the use of metabolic and genetic engineering, because scientists know the genes that are responsible for the synthesis of fatty acid in bacteria [61]. So it is easier to "produce" new bacteria using DNA recombination. For instance, the well-known *Escherichia coli* was converted into oleaginous organisms by engineering their metabolism [62]. Such genetically modified *E. coli* could produce biodiesel (fatty acid esters) directly. Kalscheuer et al. [63] obtained fatty acid ethyl esters (FAEE) with concentrations of 1.28 g L^{-1} and FAEE content in cells was 26% of the cellular dry mass during fed-batch fermentation using renewable carbon sources. Although the yield was low, it opened a new perspective for biofuel production.

We can say that microbial oil might become one of the potential feedstock for biofuels production in our future. The advantages of it are renewability, fast growth rate of microorganisms and the fact that cultivation of these organisms does not take arable lang. Use of genetic engineering and metabolic engineering can improve oil production by oleaginous microorganisms.

3. Fermentation as a process of biofuels production

Fermentation is a natural metabolic process, in which microorganisms (bacteria or yeast) obtain energy through conversion of organic compounds such as sugar and starch lipids, into simpler liquid or gaseous substances. The kind of final product depends on the metabolic pathway occurring within a cell. Some of these processes have been applied in a large industrial scale for the production of food or other valuable products. Some of these products can be used as biofuels, for example, ethanol, butanol, biomethane or hydrogen.

3.1. Alcohols

The conversion of carbohydrates (sugars and starch) with a general formula of $C_x(H_2O)_y$ from different crops, for example, potatoes, corn and cereals, into bioethanol is a common process used commercially on a large scale. Bioethanol is an oxygen compound, which is the most studied as a fuel biocomponent. This alcohol is commonly added to gasoline as a component; it can also be used as E95 ethanol fuel to supply diesel engines [64] or as a hydrogen carrier for fuel cells. Researches were also conducted to use bioethanol as a biocomponent to diesel oil [65, 66], but disadvantages of that alcohol are limited miscibility with diesel oil and high affinity to water, which influence poor physical stability of fuel.

Ethanolic fermentation is generally performed with the use of yeast such as *Saccharomyces ceveresiae*. This process consists of several stages. During first stage sucrose is hydrolysed to glucose and fructose by yeast invertase enzyme.

$$C_{12}H_{22}O_{11} + invertase \rightarrow 2\ C_6H_{12}O_6$$

Next, glucose molecule is broken down into two pyruvates (CH_3COCOO^-), which are then broken down into two acetaldehydes and CO_2. In the last stage, acetaldehyde molecules are converted into two ethanol molecules. The yeast enzyme that converts simple sugars to ethanol is called zymase, and these steps can be summarized using the following formula:

$$C_6H_{12}O_6 + zymase \rightarrow 2\ C_6H_5OH\ +\ 2\ CO_2.$$

The fermentation process is then followed by distillation and dehydration to anhydrous bioethanol. The technologies of bioethanol production from corn, potatoes or cereals are known very well.

Besides plants containing sugars or starch, lignocellulosic biomass can also be used for bioethanol production. Generally, the yield of ethanol in fermentation process depends on the ease with which the substrates can be decomposed to sugars. Starch is a biopolymer that is built from repeating glucose units and similarly to disaccharides and other oligosaccharides, is readily hydrolysed. The structure of lignocelluloses is more complex, and it is not easy to break it down into fermentable sugars. Before enzymatic hydrolysis, lignocellulosic biomass needs to be pretreated to decompose a complex matrix of cellulose, lignin and hemicelluloses [67]. This stage is complicated and needs a lot of energy. Its efficiency highly depends on the type of used pretreatment method and then influences enzymatic hydrolysis and subsequent stages [68]. The different pretreatment processes, the key factors that influence the efficiency and costs as well as advantages and disadvantages of methods of lignocellulosic biomass degradation were described in detail in literature, for example [69-71].

Ethanolic fermentation process, similarly as others biotechnological processes, can be conducted in batch, semi-continuous and continuous bioreactors. Continuous processes are more technologically advanced and have many advantages compared with the batch processes, for example, lower operation requirements, lower costs of bioreactors, better control of process and higher productivities caused by higher concentration of yeast cells. This high cell concentration can be reached by better controlling of process parameters and through some technological solutions such as immobilization techniques, recycling and recovery of microorganisms. One of the main disadvantages of the continuous processes is decrease in yeast ethanol productivity caused by long-term cultivation of cells under anaerobic conditions, and a part of substrate cannot be converted [67]. Alfenore et al. [72] indicated that aeration is an important parameter during fermentation. The presence of air increases the cell mass and ethanol productivity and causes reduction in glycerol production (the main by-product). It was stated that microaeration limited inhibiting effect of ethanol on cell growth. However, the presence of air leads to decrease in ethanol concentration.

Butanol is a four-carbon alcohol with a formula of C_4H_9OH. It is another compound produced by microorganisms that can be used as a fuel or fuel component. Due to the length of butanol's chain, it is easier to blend with higher hydrocarbons, including gasoline. It has also other advantages in comparison with ethanol such as being much less corrosive, less evaporative and having lower water solubility [73]. The results of the research conducted by Yang et al. [74] showed that when the butanol concentration in gasoline is below 20% v/v, the engine power level maintains without any engine modifications. Higher concentrations required optimization of some operational parameters of engine, for example, advancing the spark timing.

Butanol has four isomers: *n*-butanol (butan-1-ol), *sec*-butanol (butan-2-ol), isobutanol (2-methylpropan-1-ol) and *tert*-butanol (2-methylpropan-2-ol). Not all the isomers are produced by microorganisms. The *tert*-butanol is received in refinery, whereas others can be obtained in biological processes. Contrary to ethanol in the conversion of biomass to butanol not yeast are involved, but suitable strains of bacteria.

In biological processes, butanol can be produced from the same biomass as ethanol. As a substrate, it can be used plants containing carbohydrates, for example, sugar cane, sugar beets, corn and wheat. More interest is focused on feedstocks that do not compete for food such as Miscanthus, Switchgrass, wood and crop waste, algae biomass, and food processing waste.

The first process, which was enabled to obtain *n*-butanol, was acetone-butanol-ethanol (ABE) fermentation. The products of ABE fermentation are solvents such as acetone, butanol and ethanol present in the ratio of 3:6:1. In this process, strains of anaerobic bacteria from the class *Clostridium* are involved. The best-studied and widely used species is *Clostridium acetobutylicum*; other species such as *C. beijerinckii, C. aurantibutyricum* or *C. tetanomorphum* also give good butanol productivity. The particular strains are different in their ratio of end products or in a kind of end products (e.g., the isopropanol is produced in place of acetone) [75]. The kind of strains used for fermentation is dependent inter alia on a used feedstock, the required end products and their ratio, and resistance to bacteriophages. It should be remembered that *n*-butanol is very toxic for microorganisms. Concentration of this alcohol cannot exceed 12 grams per litre of a fermenting broth; higher quantity *n*-butanol will inhibit the production of this alcohol by bacterial cells, but some improvements have been made to enable the production of *n*-butanol with final content of 20 grams per litre of broth [76]. Processes of *n*-butanol separation from the broth and its purification are complex, difficult and expensive. Additionally, this process is characterized by a low conversion of glucose to butanol.

The ABE fermentation process is more complex than the production of ethanol. There are two main phases of fermentation. During the first one (acidogenesis phase), there are produced metabolites such as acetone, butyrate, hydrogen and carbon dioxide, which decreases the pH of the culture medium. During next phase (solventogenesis), the change in bacteria metabolism takes place; butanol, acetone, ethanol, H_2 and CO_2 are produced and pH of environment increases. When the glucose lacks in the culture medium (less than 10 grams per litre), the *Clostridium* produces only acids.

Besides glycolytic reactions in metabolic pathway for the production of acids and solvents, the reactions between pyruvate and butyryl-CoA take place. During the acidogenesis phase, the acetone is produced from acetyl-CoA and the butyrate from the butyryl-CoA. During the solventogenesis phase, both acetyl-CoA and butyryl-CoA are intermediates for the production of ethanol and butanol. Acetyl-CoA is the key intermediate for the synthesis of acetone. Some strains of *Clostridia,* for example, *Cl. beijerinckii* and *Cl. Aurantibutyricum,* reduce acetone to isopropanol in later stages [77]. Simplified pathway of ABE fermentation by *Clostridia* is presented on Figure 6.

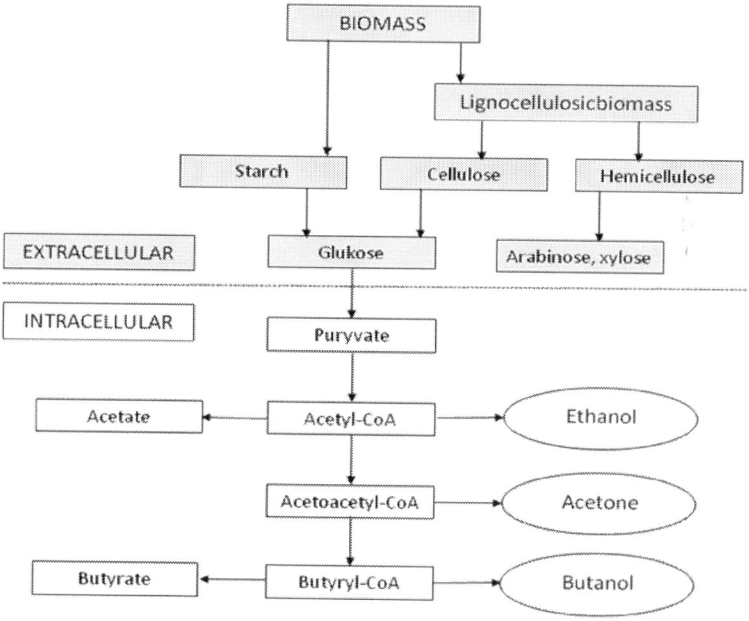

Figure 6. Simplified ABE fermentation diagram pathway (adopted from [78]).

sec-Butanol is not obtained directly by fermentation. In the first step, some strains of bacteria convert sugars (from starch, cellulose or hemicelluloses) to an intermediate product by fermentation. In the next step, this product (directly in the fermented broth) is chemically converted to *sec*-butanol. The efficiency of conversion by fermentation as well as chemical conversion of intermediate product to *sec*-butanol is very high - over 90%. One tonne of this alcohol can be obtained from three tonnes of carbohydrates; it is much more than in case of other butanol isomers [76].

Obtaining higher yields, higher productivity of butanol or others isomers of this alcohol, better butanol tolerance of bacteria involves genetic manipulation of the metabolic pathway within bacteria or DNA code. Such actions make the butanol production more attractive to the industry. Use of genetically modified microorganisms of *Ralstonia eutropha* H16 enabled the production of isobutanol from CO_2. Wild-type *Ralstonia eutropha,* in the presence of carbon and under nutrient deficiency, produces polyhydroxybutylate (PHB) in cells; this compound is a intracellular material for carbon storage. In modified strains of these bacteria, the excess carbon was redirected from PHB storage for the production of isobutanol and 3-methyl-1-butanol [79]. Another example of modified bacteria is *Clostridium pasteurianum* MBEL_GLY2. This microorganism was engineered to use glycerol as a major carbon source for butanol production. This hyper–butanol-producing strain was able to produce 10.8 g L^{-1} of butanol from 80 g L^{-1} of glycerol, whereas the native bacteria was able to produce only 7.6 g L^{-1} of butanol [80]. Progress in the efficiency of the butanol production (substrates, yield and productivity) and in the butanol separation from the broth decreases production costs and causes the production of butanol as a fuel to be more profitable.

3.2. Biogas

Methane fermentation is a well-recognized process, which is widely used for biogas production. It is one of the methods of organic waste management (e.g., livestock manure, sludge from sewage treatment plants or organic fraction of municipal waste). As a feedstock for this process, we can also use biomass from landfill or energy crops.

Biogas is produced by anaerobic microorganisms from organic substances during digestion. Methane and carbon dioxide are the reaction products, as well as small amounts of nitrogen, hydrogen and hydrogen sulphide. The process of methane fermentation is divided into several stages, conducted by several groups of interdependent microorganisms. Products of particular stages become a food for the next group of bacteria. Methane is a bacterial metabolic waste of the last fermentation step.

There are four main phases in the production of biogas: hydrolysis, acidogenesis, acetogenesis and methanogenesis - Figure 7. In the various phases, different groups of microorganisms that remain in syntrophic relationships are involved and have different environmental requirements [81].

At hydrolysis stage, the complex organic matter (e.g., carbohydrates, fats and proteins) is decomposed to simpler compounds (e.g., amino acids, sugars and fatty acids). In this process, extracellular enzymes (hydrolases) of bacteria break down the organic material by biochemical

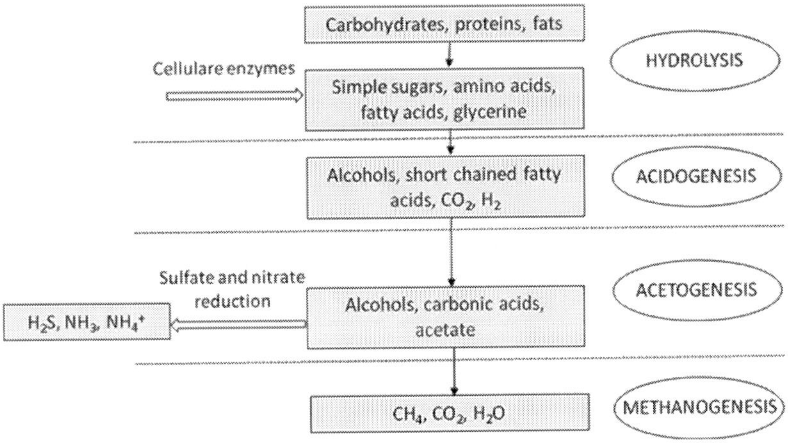

Figure 7. Simplified methane fermentation diagram.

reactions. Facultative anaerobic bacteria involved in this stage consume oxygen dissolved in water, which is usually supplied with substrate. In this way the anaerobic environment is ensured for further changes during methane fermentation. The duration of hydrolysis depends on the type of a raw material. Hydrolysis of carbohydrates takes several hours, whereas of proteins and fats takes several days. Lignocellulose and lignin are decomposed very slowly and their hydrolysis is incomplete.

During acidogenesis, the acidogenic bacteria convert the soluble products from the first stage to the lower volatile fatty acid (e.g., acetic, propionic, butyric and valeric acids), ketones, alcohols (methanol and ethanol) and gases (hydrogen and carbon dioxide). In contrast to the hydrolysis, the reactions of this phase take place within the bacterial cells. The bacteria involved in hydrolysis and acidogenesis are almost the same. They could be facultative anaerobic (e.g., *Enterobacterium* and *Streptococcus*) or obligatorily anaerobic bacteria (e.g., *Bacillus, Clostridium,* and *Bifidobactrium*).

The products of second stage are transformed by acetogenic bacteria into acetic acid, hydrogen and carbon dioxide (methanogenic substrates) at acetogenesis. Acetogenic bacteria are very sensitive to environmental changes and require long periods to adapt to new conditions. They are mandatory producers of H_2, but the growth of these bacteria is strongly inhibited by even small increase of hydrogen pressure. Acetogenic bacteria can only survive in symbiosis with genera that consume hydrogen, for example, hydrogenotrophic methanogens [81].

During the last stage (methanogenesis), methanogens, under strictly anaerobic conditions, convert the carbon dioxide, hydrogen and acetic acid to methane. Almost two-thirds of methane are produced from acetate or alcohols (acetoclastic methanogenesis) and one-third from reduction of carbon dioxide (hydrogenotrophic methanogenesis) [82].

The biogas composition depends on a type and a chemical composition of substrates, used technology and operational parameters. Methane fermentation process is extremely sensitive to changes in ambient conditions because of delicate balance between cooperating groups of involved microorganisms. There are a lot of factors that affect the biogas productivity, for example, temperature, pH, mixing, redox potential, nutrients (C/N/P ratio), inhibitors, trace elements, organic loading and hydraulic retention time [81, 83]. For example, too low C/N ratio leads to increased ammonia production and inhibition of methane production. On the other hand, too high C/N ratios cause lack of nitrogen and have a negative influence on bacterial metabolism [84]. Temperature of the process is also very important. The rate of the decomposition of organic substances increases when temperature increases till the optimal value is reached. The better the decomposition, the better the efficiency of biogas production. Increase in temperature over optimal value causes the protein denaturation and microorganisms' death; so the rapid decrease in process rate takes place. There are three temperature ranges for methanogenic fermentation: psychrophilic, mesophilic and thermophilic [85].

Generally, the biogas production systems are technologies involving one or two technological steps. In one-step process, all phases of anaerobic digestion take place in one digester. In two- or multi-step process, the biochemical phases are physically separated and are conducted in different bioreactors. Very often, fermentation process is realized using a wet digestion technology. In this technology, the dry matter content is from 12% to 15% and the feedstock can be easily pumped and stirred. If the dry matter content exceeds 16%, the culture medium loses the ability to pump; so this process is called to be dry [86].

Similar to other biotechnological processes, different types of biodegradable substrates can be used for anaerobic digestion, including lignocellulosic. When the biomass contains a large amount of lignocellulose, the microorganisms are not able to decompose it. In this case, it is important to break these structures by biomass pretreatment. The destruction of the lignocellulose structure causes the biomass to become more available for microbes, which resulted in increase in biogas and methane yield. The methods of biomass pretreatment are similar to those described above for conversion of microalgae. The biomass for methane fermentation can be prepared mechanically, physically, chemically, biologically or using mixed methods [71, 87, 88].

The next step is a biogas purification and/or upgrading process. The purification/upgrading of biogas can be realized with different methods. The choice of method of a biogas treatment is dependent on the destination of the final product. To produce a hot water or steam in boilers, biogas needs to be filtrated to remove particles and purified from steam and sulphur compounds that affect corrosion (hydrogen sulphide). Similar purification should be conducted when we want to use biogas in cogeneration (CHP process), for the simultaneous generation of usable heat and electricity. In cogeneration, a power plant from 1 m^3 of the biogas, 2.1 kWh of electricity and 2.9 kWh of heat could be generated [89]. Biogas can also be upgraded to the quality of natural gas and injected to gas grid or used as a vehicle fuel [90, 91]. In this case, in addition to removal of particulates, H_2S and water vapour, it is necessary to significantly reduce CO_2 content to meet the requirements of natural gas as Wobbe Index or calorific value. Biogas should be additionally cleaned from the trace components that are harmful to the natural gas grid or vehicle engine. For instance, siloxanes could form SiO_2 under combustion conditions, which deposits at spark plugs, valves and cylinders causing the wear of surface. Ions such as Cl^- and F^- cause corrosion of engine elements. After cleaning operations, such gaseous product contains 95–97% CH_4 and 1–3% CO_2 [91].

Purification of biogas can be realized by different methods that were described in many publications [81, 90, 92]. Separation of H_2S can be realized with the use of:

- biological methods – commonly with the use of microorganisms such as *Thiobacillus* and *Sulfolobus*, which reduce the hydrogen sulphide to the elemental sulphur and sulphates;
- physical methods as (a) precipitation of sulphides using iron(II) or (III) chloride iron (II) sulphate, (b) absorption in iron chelate solution, (c) desulphurization with bog iron ore, and (d) adsorption on an activated charcoal.

CO_2 from biogas can be removed by physical methods such as high-pressure water wash, pressure swing adsorption, chemical adsorption (e.g., in amines), precipitation of CO_2, for example, by CaO, membrane separation and cryogenic separation.

Upgraded biogas can be used as a fuel in vehicles in the similar forms as a natural gas: compressed (CNG), liquid (LNG) and adsorbed (ANG).

As mentioned earlier, one of the products produced during first stages of anaerobic digestion is hydrogen. If this part of fermentation (the so-called dark fermentation) is conducted in separated bioreactor, we can get hydrogen, in addition to methane. The amount of obtained hydrogen highly depends on hydraulic retention time, the pH value and gas partial pressure [93]. The higher yield of H_2 is obtained when the feedstock is rich in carbohydrates. The main products of the dark fermentation are H_2 and CO_2 combined with other gases, such as CH_4 or H_2S, depending on the reaction process and the used substrate. From the model substrate, which is glucose, maximum 4 mole of H_2 are produced from 1 mole of glucose when additional final product is acetic acid:

$$C_6H_{12}O_6 + 2H_2O \rightarrow 2CH_3COOH + 2CO_2 + 4H_2$$

When the end product is butyrate, the yield of hydrogen is two times smaller:

$$C_6H_{12}O_6 \rightarrow 2CH_3CH_2CH_2COOH + 2CO_2 + 2H_2$$

Generally, the end product is a mixture of both acetate and butyrate, therefore the yield of 4 moles of H_2 per 1 mole of glucose cannot be achieved [94]. This fermentative process of hydrogen production has relatively low energy conversion efficiency, so the improvement of efficiency of H_2 generation is the main challenge for researchers.

The separation of methane fermentation process into two stages has also an additional advantage. Biogas from the second step of fermentation, collected in a separate container, is richer in CH_4 because it is not diluted with CO_2, which is produced mainly in the early stages of fermentation. This indicates that with suitable plan of multi-step process and by appropriate selection of the parameters, we can obtain two valuable gaseous fuels.

3.3. Gas fermentation

Direct fermentation of biomass to produce biofuels or other chemicals is not always possible and profitable. When the biomass contains a large amount of lignocelluloses, microbial enzymes or biomass pretreatment methods are not able to break effectively this structure into simple compounds. Biomass can also be converted into biofuels with the use of microorganisms, but in an indirect way. In first step biomass is gasified to produce gaseous mixture containing CO_2, CO, H_2, CH_4 and N_2 (the so-called producer gas). After cooling and purification from tars (e.g., by stripping with solvents or by catalytic gasification of tars), obtained gas can be upgraded to syngas. Next, the syngas can be subjected to the thermochemical FT process, where it is converted to liquid fuels in the catalyst presence. Another possibility is the fermentation of purified producer gas or syngas by microorganisms leading to valuable products such as ethanol [95]. An advantage of the fermentation process in relation to the FT process is its lower cost. The costs connected with FT thermochemical route are generally generated by metal catalysts, which are expensive because of their limitations in the robustness, flexibility and selectivity [96]. Biocatalysts, which are cells of microorganisms, are cheaper than metal catalysts. They are highly specific what improves product yield and simplifies product recovery. Furthermore, biocatalysts are more resistant to sulphur contamination (presence of H_2S, carbonyl sulphide, mercaptans in producer gas) than inorganic catalyst, and the H_2:CO ratio has no influence on it [97]. This hybrid thermo/biochemical process of biomass conversion has an additional advantage – organic feedstock, which is toxic for conventional fermentative microbes, can be fermented after gasification (difference in chemical composition of feedstocks is unimportant for biomass gasification). However, some disadvantages of gas fermentation, causing that this process is not commercialized, are its low productivity and limited mass transfer between gas and liquid phases [98]. The mass transfer can be improved by reducing the mass transfer resistance at the gas–liquid interface or by reducing the surface tension, increasing the gas solubility in the liquid. This can be achieved, for example, by increasing the speed of agitation, by increasing the gas flow rate, using impeller in stirred tank reactor or by special reactor configurations (e.g., bubble column reactor, air-lift reactor, trickle-bed reactor or immobilized cell reactor) [99]. To significantly improve the mass transfer rate, the fermentation process can be carried out using reducing surface tension chemicals such as bio-polymers (xanthan gum and dextran), bio-surfactants (biological detergents) and organic compounds (high carbon alcohols and perfluorocarbon compounds) [100]. Addition of 0.1 volume percent TYLOXAPOL™ detergent causes the CO mass transfer rate to increase to over 300%.

Some microorganisms need CO and H_2 (which are present in producer gas) for metabolism; these compounds are sources of carbon and energy for bacteria. Final products are different and depend inter alia on the strain of bacteria, which can be acetate, butyrate, formate, butanol, ethanol and hydrogen [101]. Similar to methanogenic bacteria, bacteria for gas fermentation have different necessities for an optimal growth. Some species live in moderate temperatures

30-40°C (mesophilic organisms) or in higher temperatures 55-60°C (thermophilic organisms). However, it was recognized that the optimal growth of some bacteria occurs at temperatures of 70-80°C. The high temperature of fermentation process is not a problem because syngas is obtained by gasification at a very high temperature (700-800°C), and it must be cooled before introducing into the bioreactor. However, the processes of gas fermentation with the use of thermophilic bacteria are still at the research stage. Gas-fermentative organisms can differ in optimal pH value (from 5.8 to 7.5) and time when they are doubling their biomass (from 1 hour to 140 hours). Similar to other fermentations, the efficiency of gas fermentation process depends on nutrient concentration, pressure and agitation speed. The content of trace metals in the reaction medium is also very important. Some of them can enhance the cell growth and ethanol production (e.g., Zn^{2+} and Fe^{2+}), whereas some elements may have a negative impact on the process (e.g., Cu and Mo) [99]. Examples of microorganisms and final products of fermentation process are presented in Table 6.

Microorganisms species	Optimal growth conditions	Final products
	Bacteria	
Clostridium autoethanogenum	*T*=37°C, pH=5.8-6.0	Acetate, ethanol
Clostridium carboxidivorans	*T*=38°C, pH=6.2	Acetate, ethanol, butyrate, butanol
Acetobacterium woodii	*T*=30°C, pH=6.8	Acetate
Butyribacterium methylotrophicum	*T*=37°C, pH=6	Acetate, ethanol, butyrate, butanol
Rubrivivax gelatinosus	*T*=34°C, pH=6.7-6.9	Hydrogen
Moorella thermoacetica	*T*=55°C, pH=6.5-6.8	Acetate
Carboxydibrachium pacificus	*T*=70°C, pH=6.8-7.1	Hydrogen
Thermincola carboxydiphila	*T*=55°C, pH=8	Hydrogen
	Archaea	
Methanosarcina barkeri	*T*=37°C, pH=7.4	Methane
Methanothermobacter thermoautotrophicus	*T*=65°C, pH=7.4	Methane
Thermococcus strain AM4	*T*=82°C, pH=6.8	Hydrogen
Archaeoglobus fulgidus	*T*=83°C, pH=6.4	Acetate, formate, hydrogen sulphide

Table 6. Anaerobic microorganisms and final products of the fermentation process [101].

Synthesis of acetate, butyrate, ethanol and butanol by bacteria from syngas is realized via the acetyl-CoA pathway. Acetyl-CoA is produced in two major steps. During the first step, H_2 is oxidized to $2H^+$ or CO to CO_2 (with H_2O). Next, the CO_2 and $2H^+$ are reduced to formate (HCOOH), which is then converted into methyl group through a series of reactions. In the second step, the methyl, carbonyl and the CoA groups take part in the synthesis of acetyl-CoA with the use of enzymes. Further reduction of acetyl-CoA produces acetate and ethanol. The butyrate and butanol are produced by reduction of acetoacetyl-CoA that is formed from two acetyl-CoA molecules [101]. The reactions taking place during CO and CO_2 fermentation via acetyl-CoA pathway can be summarized [98] as follows:

$$6CO + 3H_2O \rightarrow C_2H_5OH + 4CO_2 \quad (1)$$

$$2CO_2 + 6H_2 \rightarrow C_2H_5OH + 3H_2O \quad (2)$$

$$4CO + 2H_2O \rightarrow CH_3COOH + 2CO_2 \quad (3)$$

$$2CO_2 + 4H_2 \rightarrow CH_3COOH + 2H_2O \quad (4)$$

The composition of syngas is a very important factor for the yield of fermentation process. For instance, during fermentation to ethanol from CO alone, from Eq. (1), one-third of the carbon from CO can be theoretically converted to alcohol. Practically, not all carbon can be converted to ethanol, because a part of CO is oxidized during metabolic pathway of bacteria to produce CO_2 and reduce equivalents. For the mixture of H_2 and CO_2 in the ratio 3:1 (Eq. 2), theoretically all carbon can be converted to C_2H_5OH, but it is not possible to obtain this ratio of mentioned gases during gasification of biomass. H_2 molecule supplies H^+ ions and electrons required in reactions of hydrogenase enzyme, leading to ethanol synthesis. Thus, in this case, a part of H_2 is used for the production of reducing equivalents [102].

Gas obtained from biomass contains different components, for example, gaseous compounds containing sulphur and nitrogen, tar, ethylene, acetylene and ash. They can affect the efficiency of gas fermentation, inhibiting the microbial catalyst, what influences the product yield. For example, the contamination of syngas by NO in concentrations above 40 ppm causes inhibition of hydrogenase enzyme and the cells of bacteria stopped consuming H_2. It changes the final product composition because CO is used in place of hydrogen for electron production rather than in product formation [103]. Similarly, acetylene is a strong inhibitor of hydrogenase and also affects the hydrogen consumption by cell [102]. Additionally, research results obtained by Datar et al. [102] indicated that production of ethanol is associated with growth of bacteria during producer gas fermentation. Ethanol was primarily produced once the cells stop growing.

Content of sulphur and nitrogen oxides in the producer gas can be reduced by the removal of N- and S- compounds from biomass. Reduction of the concentration of some elements in biomass can be achieved by its pretreatment. Turn et al. [104] studied fuel characteristics of sugarcane, which were subjected to a single milling and an initial milling, followed by leaching and a secondary milling. They stated that milling caused reduction of ash. They explained it by removal of K, Cl, S and N in the treatment operations. The second pretreatment method (milling combined with leaching) was more effective in the removal of the mentioned elements than the first one.

Some products can have an inhibiting effect. Najafpour et al. [105] indicated that during the batch fermentation of synthetic mixed gas (55% CO, 10% CO_2, 20% H_2 and 15% Ar) by *Rhodospirillum rubrum* ATCC 25903, the increase in the acetate content to 3g/L caused significant decrease in the conversion of acetate and CO and led to the reduction of H_2 production.

The yield of gas fermentation products depends on both the kind of biomass and the strain of microorganisms. Liu et al. [106] stated that replacing yeast extract with corn steep liquor in batch fermentation with the use of *Alkalibaculum bacchi* strain CP15 leads to increase in ethanol production by 78%. They also reported that fermentation of yeast extract under continuous conditions enabled to achieve C_2H_5OH concentration of 6 g/L, but when the corn steep liquor was a substrate, the maximum produced concentrations of ethanol, n-propanol and n-butanol were 8 g/L, 6 g/L and 1 g/L, respectively. High cell mass concentration above 5 g/L decreases ethanol production. Some microorganisms are known to convert syngas to ethanol, but very few can produce higher alcohols alone. The solution could be fermentation by mixed culture of bacteria [107]. Using *Alkalibaculum bacchi* and *Clostridium propionicum* to syngas fermentation results in over 60% more alcohol yield than with A. *bacchi* CP15 alone. Moreover, the mixed culture converted 50% more carboxylic acids (propionic acid, butyric acid and hexanoic acid) into their corresponding alcohols than the CP15 monoculture. The alcohol concentration can be increased also by appropriate composition of liquid phase. Najafpour and Younesi [108] examined the syngas fermentation using batch culture of *Clostridium ljungdahlii*. It was stated that initial presence of hydrogen and carbon dioxide in the liquid phase enhanced the ethanol concentration twice.

Main disadvantages of gas fermentation such as sterile conditions, slow reaction rate and mass transfer limitations cause that there is a lack of commercial plants. Most installations exist in the laboratory or pilot scale. Genetic engineering can help in faster progress regarding the implementation of this process by improving the sensitivity of involved microorganisms to the high concentrations of end products as well as the efficiency of the syngas conversion to valuable chemicals.

4. Energy and fuel production with microbial BESs

Many microorganisms are naturally capable of transporting electrons in and out of the cell – electrogens. In nature, this phenomenon is used for, for example, mineral reduction, but it can be exploited for harvesting electricity from or to provide electricity to microbial communities [109]. All bioelectrochemical systems (BESs) cosist of anode, where the oxidation reaction occurs, and cathode, where the reductive reaction takes place. At least one of these processes is catalysed by microorganisms (microbial electrocatalysis), and it brings on the terms microbial bioanode and biocathode [110]. In bioanodes, exoelectrogenic bacteria anaerobically oxidize organic or inorganic matter, discharge electrons and transport them through the electron transport chain to the electrode surface directly or via mediators. The direct electron transport occurs by a contact with the electrode surface through conductive proteins that are the integral elements of the microorganism cell membrane (e.g., cytochrome). The indirect electron transfer is realized through substances with redox properties that act as electron carriers and transport them from cell membrane to the anode surface [111]. In biocathodes, electrotrophs collect electrons from the cathode (directly or using mediators) and reduce different compounds such as organics, carbon dioxide, sulphate or nitrate [109, 110]. There is a wide range of BESs that have been developed for different processes such as power generation (microbial fuel cells, MFC), biofuels and biochemical production (BES), waste remediation (bioelectrochemical treatment systems BET), production of H_2 at low applied potential (microbial electrolysis cells, MEC) and others. The schemes of exemplary BES configurations are presented in Figure 8. The most common process based on microbial electrocatalysis is electricity production in MFCs [112, 113]. BESs can be inoculated with a wide spectrum of bacteria, for example, *Shewanella oneidensis, Geobacter sulfurreducens, Moorella thermoacetica, Clostridium ljungdahlii, Escherichia coli* or *Acetobacterium woodii* [114]. Microorganisms can be used as a monoculture or as a culture mixture.

MFC devices transform the chemical energy in organic matter into electrical power. Because the chemical energy from the oxidation processes is directly turned into the electricity instead of heat, the Carnot cycle with limited thermal efficiency can be omitted, as it is in conventional chemical fuel cells (more than 70%) [115]. MFCs, as this is a new technology, have tremendous potential because of their operational and functional advantages [115]:

- high conversion efficiency of direct conversion of energy contained in organic or inorganic substrate to electricity,
- efficient operation in ambient and even low temperatures (unlike the all other current bioenergy generation processes),
- the gaseous off-products of the process do not require the treatment,
- no energy input for aeration needed (passive aeration),
- applications in locations without the electrical infrastructure.

The selection of MFC apparatus components and design, along with the microbial consortia, is crucial for the performance of whole device and is limited by the specific requirements posed by the nature of the electrochemical processes and the bacteria. All materials that are used for MFC construction should be chemically stable, biocompatible and insensitive for impurities and additionally resistant to corrosion. What is more is that the materials for electrodes should have large active surface area, high porosity and conductivity. The structure of their surface should not impair the electron transport process. Of course, it is desired that they should be simultaneously cheap and easy to manufacture. The search for the perfect electrode material is still on-going – there is a great hope in graphene. So far, the most commonly used materials are carbon and graphite paper, cloth, fibres and meshes, reticulated vitreous carbon and stainless steel [113, 116].

Figure 9A presents the power density values (referred to anode active surface area) that were obtained in different MFC configurations. As it can be seen, the differences reach more than two orders of magnitude. Figure 9B presents the maximal values generated in MFC voltage and maximal power densities (referred to active cathode surface area) for different cathodes.

http://dx.doi.org/10.5772/62397

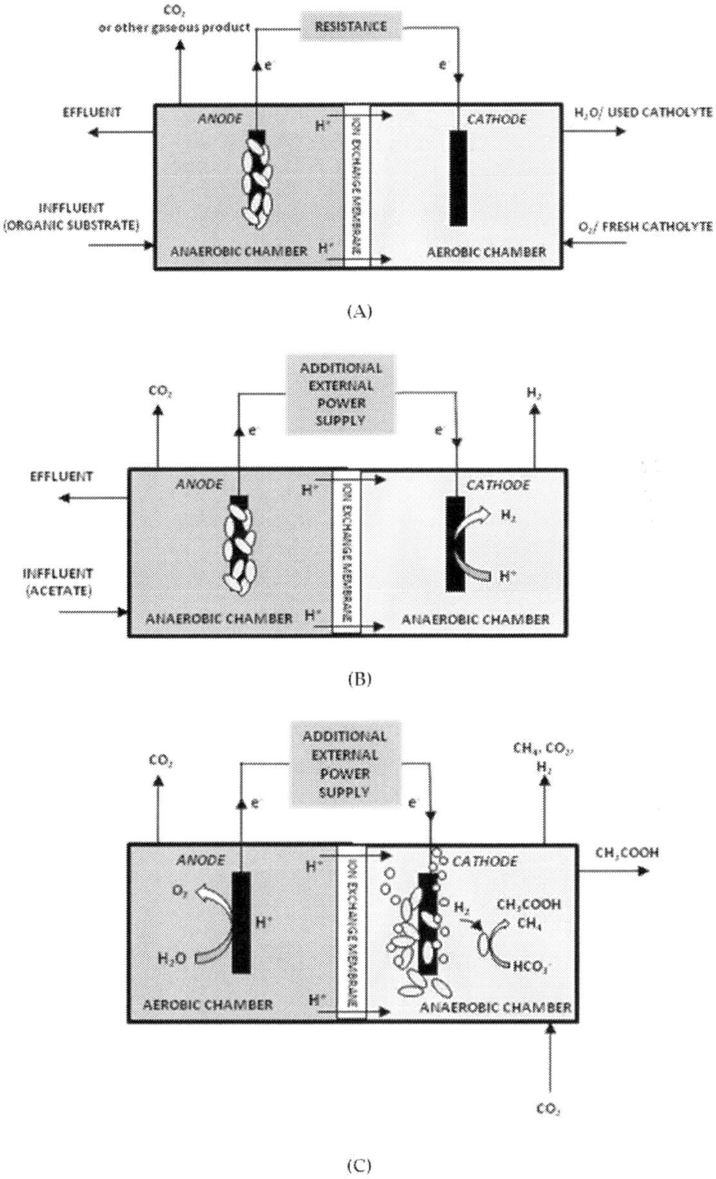

Figure 8. Schematic diagrams of exemplary two-chamber BESs configurations: A – microbial fuel cell (MFC), B – microbial electrolysis cell (MEC), C - electrosynthesis microbial cell (MES).

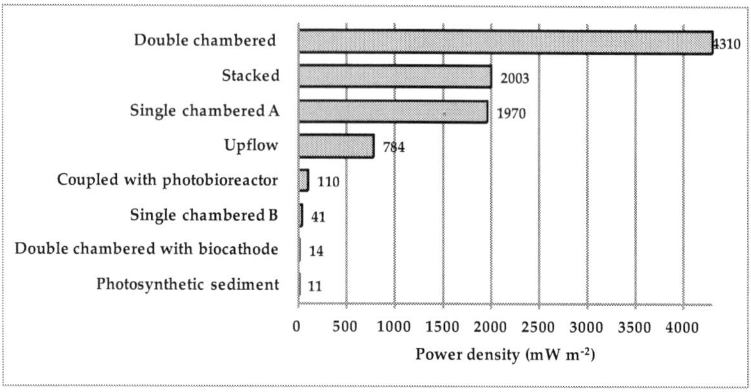

(A)

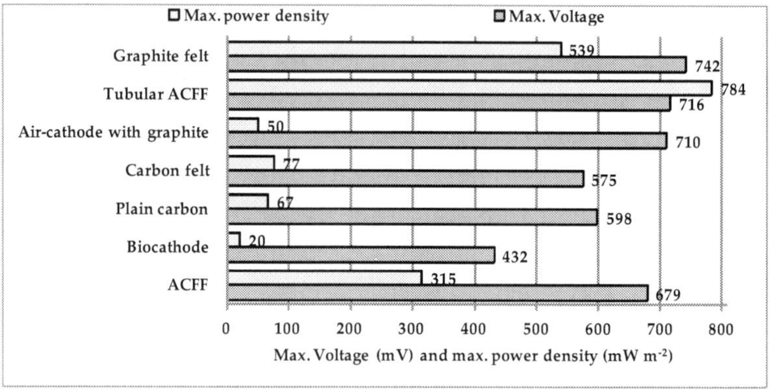

(B)

Figure 9. A – representative power densities obtained from different configurations of MFC (adopted from [112]), B – maximum generated voltages and maximum power densities for different cathodes (adopted from [117]).

This clearly shows that change in even one MFC component has potentially enormous influence on the whole system efficiency.

MFCs enabled electricity production from biodegradable raw materials, for example, wastewater. They seem to be an alternative to expensive conventional aerated wastewater treatment. Huggins et al. [118] compared traditional aeration process in wastewater treatment plant with MFC. They indicated that MFC showed lower removal efficiency when the carbon oxygen demand (COD) concentration is high, but it is much more effective than aeration when the COD is low. It also significantly reduces sludge production (by 52–82% as compared with

http://dx.doi.org/10.5772/62397

aeration), which can reduce the size of secondary clarifier and save the cost of sludge disposal. Furthermore, MFCs save 100% of aeration energy with extra electricity output.

Another application is that BESs uses them for underground contaminant remediation where the electrodes are an inexhaustible source of terminal electron acceptors for a groundwater environment. Process can be conducted with the use of a single or array electrodes without using enclosed containers. In such a solution, bacteria are simulated to decomposition of underground pollutants and produce additional electricity. It eliminates chemicals, which are indispensable in conventional technologies and reduces cost of energy [119]. Remediating MFC technology can be used, for example, for removal of petroleum hydrocarbons by their oxidation. Morris et al. [120] indicate that anaerobic biodegradation of petroleum derivatives was importantly enhanced in an MFC (82% removal) in comparison with an anaerobically incubated control cell (31% removal) over 21 days at 30°C [120]. When the electrode is used as an electron donor, the oxidized contaminants, such as chromium, perchlorate, chlorinated solvents and uranium, can be reduced using the electrode as the electron donor [119].

MFCs can be easily adjusted to the generation of hydrogen, instead of producing electricity. Produced H_2 can be accumulated for different applications. When the anodic potential in the MFC device increases using an external voltage of about 0.23 V or more, the gaseous hydrogen is produced at the cathode by the reduction of protons (in the absence of oxygen in the cathode chamber) [117]. It is the so-called MEC. The main advantages of such a hydrogen production are [119, 121]:

- used external power is much lower (0.2 V) than used in traditional water electrolysis (1.23 V),
- no expensive catalysts needed on the anode,
- waste and renewable materials can be used as a substrates,
- the H_2 production rate can be three times higher than during dark fermentation.

Despite the big potential of microbial electrolytic production of hydrogen, the development of this technique is still at the laboratory scale. Successful application of the MEC is possible when the problems with low efficiency and with scalability of device are resolved.

In case of two-chamber devices, the chambers are separated by a membrane that is expensive and can increase the internal resistance at pH 7.0. The membrane also caused large potential losses associated with the exchange of anions and cations. It resulted in considerable increase in the value of applied voltage and decrease in the energy recovery [121]. On the other hand, one-chamber MECs have simpler construction (membrane-free), lower internal resistance and much higher yield of H_2, but produced hydrogen is consumed by methanogenesis to generate methane [119, 121].

MEC systems can be used for the production of other inorganic compounds in the cathode chamber. For example, Rozendal et al. [122] stated that phosphate can be recovered as struvite in a modified MEC. Single-chambered reactors are characterized by the low efficiency of phosphorus removal from wastewater (20–50%), and there is a problem with failure of cathode because of scale accumulation [122]. Rozendal et al. [122] designed a two-chamber MEC with a fluidized bed to decrease scale formation on the cathode surface. This reactor promotes bulk phase struvite precipitation and protect electrode. Additionally, it has a high level of soluble phosphorus removal (70–85%), compared to 10–20% for the control (open circuit conditions). The method of struvite recovery with the use of MEC device is consuming less energy than conventional pH adjustment systems (e.g., aeration and chemical base addition). MEC also enables a production of hydrogen peroxide (H_2O_2) by reduction of oxygen at a cathode [123]. It was possible to obtain ~1.9 ± 0.2 kg H_2O_2/m^3/day from acetate at an overall efficiency of 83.1 ± 4.8%, where the applied voltage was 0.5 V.

Microbial electrosynthesis systems (MES) enabled the production of organic compounds and fuels by reduction of CO_2 or other chemicals on the cathode. The concept of MES is quite new (2009–2010); the devices were elaborated when it was discovered that electrical current can change microbial metabolism. One of the generated products is acetate, which can be converted to liquid fuels, but this process is characterized by low rates and yields. Generally, acetogenic bacteria can reduce CO_2 to acetate using hydrogen as a donor of electrons. It was found that some acetogens, for example, *Geobacter* and *Anaeromyxobacter* species, accept electrons from graphite electrodes for chemical productions in place of hydrogen electron donor [124]. Therefore, Nevin et al. [124] investigated the possibility of carbon dioxide reduction to acetate using acetogenic microorganism *Sporomusa ovata*, where the electrons were delivered directly to the cells with a graphite electrode. They stated that the conversion of CO_2 and H_2O to an organic compound and oxygen:

$$2CO_2 + \ 2\,H_2O \rightarrow CH_3COOH\ 2O_2$$

is much similar to the reaction taking place during oxygenic photosynthesis. Marshall et al. [125] improved acetate yield by operating established biocathode in semi-batch mode at a potential of –590 mV (versus SHE) for over 150 days. Maximum acetate production from CO_2 (as the only carbon source) was 17.25 mM day^{-1} (1.04 g L^{-1} d^{-1}). Beside acetate, the second main product was hydrogen. Steinbusch et al. [126] showed that obtained acetate can then be converted into ethanol by biological reduction with the use of electrode, in hydrogen stead. To stimulate acetate reduction at the cathode with the mixture of bacteria cultures, the addition of a mediator (methyl viologen) was required. At applied cathode potential –550 mV, used mediator enhanced ethanol production 6-fold and increased ethanol concentration 2-fold compared with the control probe. Generally, the microbial electrosynthetic cells have great potential, especially in biofuel production, but there are still many technological as well as economical challenges to be solved before their implementation in industrial scale.

From among microbial BESs, the most advanced research is in the field of MFC, but there are also a lot of challenges facing researchers. Similarly as in other systems, there is a problem with upscale of MFC. With increasing size of MFC, the power density generally decreases. In bigger MFC devices, the distance between anode and cathode electrodes is also bigger, which influences increase in resistance and pH slope of solution [112]. The MFC can be connected parallel or in series to increase produced current and voltage, but there are also some problems, especially when any cell is weak. The other problem connected with upscaling is high cost of membrane electrodes, which additionally should be resistant to degradation by microorganisms. It is also important to for example [112]:

- find cheaper materials for construction of reactors, electrodes, membranes etc.,
- improve metabolism of microorganisms for better extracellular transfer of electrons,
- improve catalytic efficiency of microorganisms,
- improve electrolyte conductivity,

and others. However, in the near future, the microbial BESs could revolutionize the market for the production of sustainable energy, fuels and chemicals.

5. Conclusions

Technological progress is inextricably linked to the increase in demand for energy. Rapid development of industrial and automotive sectors caused depletion of the deposits of fossil fuels, which resulted in the necessity of searching for alternative sources of energy. The microorganisms such as bacteria, microalgae, fungi and yeast can become allies in meeting the growing needs of people for fuels. These organisms can produce valuable substances (used as biofuels or as a substrates for their production) as a product of organism's metabolism or can be a raw material for technologies converting their biomass into biofuels.

There are many ways in which fuels can be obtained with the use of microorganisms. At present, the greatest hopes are set on technologies that are well known as ethanolic fermentation or anaerobic digestion of biodegradable organic wastes to biogas. Researches in the area of these technologies are focused mainly on improving the efficiency of the process, the search for new substrates or on the methods of pretreatment of raw materials. Promising microorganisms seem to be unicellular algae. These organisms are source of different valuable compounds such as pigments, lipids, sterols, fatty acids, starch, oils and others. Microalgae are both producers and raw materials for biofuel production. Their conversion to energy can be realized in different biochemical as well as thermochemical routes. Microbial oil obtained from algae can be a substrate for the production of biodiesel or hydrocarbons. Also algae biomass (raw or after oil extraction) can be converted into fuels by different fermentations, gasification and liquefaction. At present, we know a lot about their metabolism, cultivation, harvesting and some technologies of their processing, but there are still a lot of problems demanding solution for improving the efficiency of microalgae cultivation, cost reduction, increasing oil content in the cells and others. Microalgae have additional advantages; they have a very big potential for the purification of water from nitrogen and phosphorus. These elements are necessary to build algae's cells and due to the enormous growth, microalgae need very large amounts of N and P. Algae can also accumulate other elements dissolved in water, which may increase the scope of their application. Furthermore, algae are photosynthetic organisms, so they need significant amounts of CO_2 during growth. It caused that these organisms can be used to capture carbon dioxide from different gases (biogas, producer gas and exhaust gas). The above-mentioned advantages of algae indicate that these organisms can be used multi-faceted. They can be a valuable source of biochemicals, may be used in the purification processes of water or gases or to recover some elements from water. Microalgae seem to be an ideal raw material for biorefineries.

Application of other microorganisms (bacteria and fungi) in fuels and energy production for example direct hydrocarbon production by bacteria or microbial BESs, seems to be very interesting issue, however, in the nearest future these technologies probably will not be implemented in a larger scale. The researches in the field of metabolism of these organisms, efficiency of valuable compounds production, systems for cultivation or for energy production are in their infancy and require a lot of time before they become commercially viable.

Author details

Anna Matuszewska[1,2*]

Address all correspondence to: a.matuszewska@pimot.eu

1 Automotive Industry Institute, Warsaw, Poland

2 Cardinal Stefan Wyszynski University, Warsaw, Poland

References

[1] Demirbas A., Demirbas F. *Algae Energy. Algae as a New Source of Biodiesel*. Springer, London, 2010.

[2] Bleeke F. et al. Isolation and characterization of new temperature tolerant microalgal strains for biomass production. *Energies*, 2014, 7, 7847-7856.

[3] Mata T. et al. Microalgae for biodiesel production and other applications: A review. *Renewable and Sustainable Energy Reviews*, 2010, 14, 217-232.

[4] Metting F. Biodiversity and application of microalgae. *Journal of Industrial Microbiology*, 1996, 17, 477-489.

[5] Harun R. et al. Bioprocess engineering of microalgae to produce a variety of consumer products. *Renewable and Sustainable Energy Reviews*, 2010, 14, 1037-1047.

[6] Tsukahara K., Sawayama S. Liquid fuel production using microalgae. *Journal of the Japan Petroleum Institute*, 2005, 48, 251-259.

[7] Amin S. Review on biofuel oil and gas production processes from microalgae. *Energy Conversion and Management*, 2009, 50, 1834-1840.

[8] Rashid N. et al. Current status, issues and developments in microalgae derived biodiesel production. *Renewable and Sustainable Energy Reviews*, 2014, 40, 760-778.

[9] Chisti Y. Biodiesel from microalgae. *Biotechnology Advances*, 2007, 25, 294-306.

[10] Sialve B. et al. Anaerobic digestion of microalgae as a necessary step to make microalgal biodiesel sustainable. *Biotechnology Advances*, 2009, 27, 409-416.

[11] Gouveia L., Oliveira A. Microalgae as a raw material for biofuel production. *Journal of Industrial Microbiology and Biotechnology*, 2009, 36, 269-274.

[12] Aatola H. et al. Hydrotreated vegetable oil (HVO) as a renewable diesel fuel: Trade-off between NOx, particulate emission, and fuel consumption of a heavy duty engine. SAE Paper, 2008-01-2500.

[13] Murata K. et al. Production of synthetic diesel by hydrotreatment of Jatropha oils using Pt–Re/H-ZSM-5 catalyst. *Energy Fuels*, 2010, 24 (4), 2404-2409.

[14] Maher K.D., Bressler D.C. Pyrolysis of triglyceride materials for the production of renewable fuels and chemicals. *Bioresource Technology*, 2007, 98, 2351-2368.

[15] Lima D.G. et al. Diesel-like fuel obtained by pyrolysis of vegetable oils. *Journal of Analytical and Applied Pyrolysis*, 2004, 71, 987-996.

[16] Belarbi E.H. et al. A process for high yield and scaleable recovery of high purity eicosapentaenoic acid esters from microalgae and fish oil. *Enzyme and Microbial Technology*, 2000, 26, 516–529.

[17] Halleux H. et al. Comparative life cycle assessment of two biofuels ethanol from sugar beet and rapeseed methyl ester. *International Journal of Life Cycle Assessment*, 2008, 13, 184–190.

[18] Dębowski M. et al. Algae biomass as an alternative substrate in biogas production technologies–Review. *Renewable and Sustainable Energy Reviews*, 2013, 27, 596-604.

[19] Samson R., LeDuy A. Biogas production from anaerobic digestion of Spirulina maxima algal biomass. *Biotechnology and Bioengineering*, 1982, 24, 1919-1924.

[20] Samson R., LeDuy A. Detailed study of anaerobic digestion of Spirulina maxima algal biomass. *Biotechnology and Bioengineering*, 1986, 28, 1014-1023.

[21] Okuda K. Structure and phylogeny of cell coverings. *Journal of Plant Research*, 2002, 115, 283-288.

[22] Gerken H.G. et al. Enzymatic cell wall degradation of *Chlorella vulgaris* and other microalgae for biofuels production. *Planta*, 2013, 237, 239-253.

[23] Geider R.J., La Roche J. Redfield revisited: Variability of C:N:P in marine microalgae and its biochemical basis. *European Journal of Phycology*, 2002, 37, 1-17.

[24] Zhong W. et al. Enhanced methane production from Taihu Lake blue algae by anaerobic co-digestion with corn straw in continuous feed digesters. *Bioresource Technology*, 2013, 134, 264-270.

[25] Yen H.W., Brune D.E. Anaerobic co-digestion of algal sludge and waste paper to produce methane. *Bioresource Technology*, 2007, 98, 130-134.

[26] Mahdy A. et al. Algaculture integration in conventional wastewater treatment plants: Anaerobic digestion comparison of primary and secondary sludge with microalgae biomass. *Bioresource Technology*, 2015, 184, 236-244.

[27] Olsson J. et al. Co-digestion of cultivated microalgae and sewage sludge from municipal waste water treatment. *Bioresource Technology*, 2014, 171, 203-210.

[28] Trivedi J. et al. Algae based biorefinery – How to make sense? *Renewable and Sustainable Energy Reviews*, 2015, 47, 295-307.

[29] Brennan L., Owende P. Biofuels from microalgae—A review of technologies for production, processing, and extractions of biofuels and co-products. *Renewable and Sustainable Energy Reviews*, 2010, 14, 557-577.

[30] Rojan P.J. et al. Micro and macroalgal biomass: A renewable source for bioethanol. *Bioresource Technology*, 2011, 102, 186-193.

[31] Spolaore P. et al. Review: Commercial application of microalgae. *Journal of Bioscience and Bioengineering*, 2006, 101, 87-96.

[32] Harun R., Danquah M.K. Influence of acid pre-treatment on microalgal biomass for bioethanol production. *Process Biochemistry*, 2011, 46, 304-309.

[33] Harun R. et al. Microalgal biomass as a fermentation feedstock for bioethanol production. *Journal of Chemical Technology and Biotechnology*, 2010, 85, 199-203.

[34] Harun R., et al. Analysis of process configurations for bioethanol production from microalgal biomass. In: Shaukat S, editor. Progress in Biomass and Bioenergy Production. InTech, 2011, 395-409.

[35] Hirano A. et al. CO_2 fixation and ethanol production with microalgal photosynthesis and intercellular anaerobic fermentation. *Energy*, 1997, 22, 137-142.

[36] Ueno Y. et al. Ethanol production by dark fermentation in the marine green alga, *Chlorococcum littorale*. *Journal of Fermentation and Bioengineering*, 1998, 86, 38-43.

[37] Deng M., Coleman J.R. Ethanol synthesis by genetic engineering in cyanobacteria. *Applied and Environmental Microbiology*, 1999, 65, 523-528.

[38] Bharathiraja B. et al. Aquatic biomass (algae) as a future feed stock for bio-refineries: A review on cultivation, processing and products. *Renewable and Sustainable Energy Reviews*, 2015, 47, 634-653.

[39] Melis A., Happe T. Hydrogen production. Green algae as a source of energy. *Plant Physiology*, 2001, 127, 740–748.

[40] Kruse O., Hankamer B. Microalgal hydrogen production. *Current Opinion in Biotechnology*, 2010, 21, 238-243.

[41] McKendry P. Energy production from biomass (part 3): Gasification technologies. *Bioresource Technology*, 2002, 83, 55-63.

[42] Hirano A. et al. Temperature effect on continuous gasification of microalgal biomass: Theoretical yield of methanol production and its energy balance. *Catalysis Today*, 1998, 45, 399-404.

[43] López-González D. et al. Comparison of the steam gasification performance of three species of microalgae by thermogravimetric–mass spectrometric analysis. *Fuel*, 2014, 134, 1-10.

[44] Elliott D. et al.: Chemical processing in high-pressure aqueous environments. 7. Process development for catalytic gasification of wet biomass feedstocks. *Industrial and Engineering Chemistry Research*, 2004, 43, 1999-2004.

[45] Brandenberger M. et al. Producing synthetic natural gas from microalgae via supercritical water gasification: A techno-economic sensitivity analysis. *Biomass and Bioenergy*, 2013, 51, 26-34.

[46] Cherad R. et al. Macroalgae supercritical water gasification combined with nutrient recycling for microalgae cultivation. *Environmental Progress and Sustainable Energy*, 2013, 32, 902-909.

[47] Minowa T. et al. Oil production from algal cells of *Dunaliella tertiolecta* by direct thermochemical liquefaction. *Fuel*, 1995, 74, 1735-1738.

[48] Guo Y. et al. A review of bio-oil production from hydrothermal liquefaction of algae. *Renewable and Sustainable Energy Reviews*, 2015, 48, 776-790.

[49] Galafassi S. et al. Lipid production for second generation biodiesel by the oleaginous yeast *Rhodotorula graminis*. *Bioresource Technology*, 2012, 111, 398-403.

[50] Amaretti A. et al. Single cell oils of the cold-adapted oleaginous yeast *Rhodotorula glacialis* DBVPG 4785. *Microbial Cell Factories*, 2010, 9, 73.

[51] Papanikolaou S., Aggelis G. Modeling lipid accumulation and degradation in *Yarrowia lipolytica* cultivated on industrial fats. *Current Microbiology*, 2002, 46, 398-402.

[52] Papanikolaou S. et al. *Yarrowia lipolytica* as a potential producer of citric acid from raw glycerol. *Journal of Applied Microbiology*, 2002, 92, 737-744.

[53] Meng X. et al. Biodiesel production from oleaginous microorganisms. *Renewable Energy*, 2009, 34, 1-5.

[54] Khot M. et al. Single cell oil of oleaginous fungi from the tropical mangrove wetlands as a potential feedstock for biodiesel. *Microbial Cell Factories*, 2012, 11, 71

[55] Beopoulos A. et al. Review. *Yarrowia lipolytica* as a model for bio-oil production. *Progress in Lipid Research*, 2009, 48, 375-387.

[56] Kosa M., Ragauskas A.J. Lipids from heterotrophic microbes: Advances in metabolism research. *Trends in Biotechnology*, 2011, 29, 53-61.

[57] Tsigie Y. et al. Maximizing biodiesel production from *Yarrowia lipolytica* Po 1g biomass using subcritical water pretreatment. *Bioresource Technology*, 2012, 111, 201-207.

[58] Dai C. et al. Biodiesel generation from oleaginous yeast *Rhodotorula glutinis* with xylose assimilating capacity. *African Journal of Biotechnology*, 2007, 6, 2130-2134.

[59] Leiva-Candia D. et al. The potential for agro-industrial waste utilization using oleaginous yeast for the production of biodiesel. *Fuel*, 2014, 123, 33-42.

[60] Gouda M. et al. Single cell oil production by *Gordonia sp.* DG using agro-industrial wastes. *World Journal Microbiology and Biotechnology*, 2008, 24, 1703-1711.

[61] Li O. et al. Perspectives of microbial oils for biodiesel production. *Applied Microbiology and Biotechnology*, 2008, 80, 749-756.

[62] Rude M., Schirmer A. New microbial fuels: A biotech perspective. *Current Opinion in Microbiology*, 2009, 12, 274-281.

[63] Kalscheuer R. et al. Microdiesel: Escherichia coli engineered for fuel production. *Microbiology*, 2006, 152, 2529-2536.

[64] Chłopek Z. Ecological aspects of using bioethanol fuel to power combustion engines. *Ekspolatacja i Niezawodność*, 2007, 3, 65-68.

[65] Satgé de Caro P. et al. Interest of combining an additive with diesel – Ethanol blends for use in diesel engines. *Fuel*, 2001, 80, 565-574.

[66] Behdad S. et al. Experimental investigation of the tractor engine performance using diesohol fuel. *Applied Energy*, 2014, 114, 874-879.

[67] Sanchez O., Cardona C. Trends in biotechnological production of fuel ethanol from different feedstocks. *Bioresource Technology*, 2008, 99, 5270–5295.

[68] Wyman C., Goodman B. Biotechnology for production of fuels, chemicals, and materials from biomass. *Applied Biochernistry and Biotechnology*, 1993, 39/40, 41-59.

[69] Alvira P. et al. Pretreatment technologies for an efficient bioethanol production process based on enzymatic hydrolysis: A review. *Bioresource Technology*, 2010, 101, 4851-4861.

[70] Yang B., Wyman C. Pretreatment: The key to unlocking low-cost cellulosic ethanol. *Biofuels, Bioproducts and Biorefining*. 2008, 2, 26-40.

[71] Hendriks A., Zeeman G. Pretreatments to enhance the digestibility of lignocellulosic biomass. *Bioresource Technology*, 2009, 100, 10-18.

[72] Alfenore S. et al. Aeration strategy: A need for very high ethanol performance in Saccharomyces cerevisiae fed-batch process. *Applied Microbiology and Biotechnology*, 2004, 63, 537-542.

[73] Wu M. et al. Life-cycle assessment of corn-based butanol as a potential transportation fuel. Argonne National Laboratory, 2007, ANL/ESD/07-10.

[74] Yang J. et al. Dyno test investigations of gasoline engine fueled with butanol-gasoline blends. SAE Technical Paper 2009-01-1891, 2009.

[75] Jones D., Woods D. Acetone-butanol fermentation revisited. *Microbiological Reviews*, 1986, 50, 484-524.

[76] Nigam P., Singh A. Production of liquid biofuels from renewable resources. *Progress in Energy and Combustion Science*, 2011, 37, 52-68.

[77] Leja K. et al. Przemysłowe wykorzystanie bakterii z rodzaju *Clostridium*. *Postępy Mikrobiologii*, 2014, 53, 15–24.

[78] Ezeji T. et al. Bioproduction of butanol from biomass: From genes to bioreactors. *Current Opinion in Biotechnology*, 2007, 18, 220-227.

[79] Lu J. et al. Studies on the production of branched-chain alcohols in engineered *Ralstonia eutropha*. *Applied Microbiology and Biotechnology*, 2012, 96, 283-297.

[80] Malaviya A. et al. Continuous butanol production with reduced by-products formation from glycerol by a hyper producing mutant of *Clostridium pasteurianum*. *Applied Microbiology and Biotechnology*, 2012, 93, 1485-1494.

[81] Deublein D., Steinhauser A. Biogas from waste and renewable resources, Wiley-VCH Verlag GmbH & Co. KGaA, Weinheim, 2008.

[82] Myinta M. et al. Anaerobic fermentation of cattle manure: Modeling of hydrolysis and acidogenesis. *Water Research*, 2007, 41, 323–332.

[83] Chen Y. et al. Inhibition of anaerobic digestion process: A review, *Bioresource Technology*, 2008, 99, 4044–4064.

[84] Ituen E., Etim G. Carbon-nitrogen ratio as a basis for the selection of substrates for optimum biogas production. *Advances in Science and Technology*, 2012, 6, 33-38.

[85] van Lier J. et al. High-rate anaerobic wastewater treatment under psychrophilic and thermophilic conditions. *Water Science and Technology*, 1997, 35, 199-206.

[86] Głodek E. et al. Pozyskiwanie i energetyczne wykorzystanie biogazu rolniczego: Proces technologiczny cz. I (*Production of agricultural biogas and its energetic use. Tech-*

nological process. Part I), Instytut Mineralnych Materiałów Budowlanych, Opole, 2007 (in Polish).

[87] Turick C. et al. Methane fermentation of woody biomass. *Bioresource Technology*, 1991, 37, 141-147.

[88] Kumar P. et al. Methods for pretreatment of lignocellulosic biomass for efficient hydrolysis and biofuel production. *Industrial and Engineering Chemistry Research*, 2009, 48, 3713-3729.

[89] Gabryszewska M. et al. Opracowanie bazy danych z zakresu biogazu i instalacji biogazowych z kogeneracją (*Development of database in the field of biogas and cogeneration biogas systems*). PIMOT, DS/128, Warszawa, 2010 (in Polish).

[90] Ryckebosch E. et al. Techniques for transformation of biogas to biomethane. *Biomass and Bioenergy*, 2011, 35, 1633-1645.

[91] Budzianowski W. Sustainable biogas energy in Poland: Prospects and challenges. *Renewable and Sustainable Energy Reviews*, 2012, 16, 342-349.

[92] Osorio F., Torres J. Biogas purification from anaerobic digestion in a wastewater treatment plant for biofuel production. *Renewable Energy*, 2009, 34, 2164-2171.

[93] Ni M. et al. An overview of hydrogen production from biomass. *Fuel Processing Technology*, 2006, 87, 461-472.

[94] Hawkes F. et al. Sustainable fermentative hydrogen production: Challenges for process optimization. *International Journal of Hydrogen Energy*, 2002, 27, 1339-1347.

[95] Wilkins M., Atiyeh H. Microbial production of ethanol from carbon monoxide. *Current Opinion in Biotechnology*, 2011, 22, 326-330.

[96] Daniell J. et al. Commercial biomass syngas fermentation. *Energies*, 2012, 5, 5372-5417.

[97] Marshall D. et al. Reactor design issues for synthesis-gas fermentations. *Biotechnology Progress*, 1999, 15, 834-844.

[98] Munasinghe P., Khanal S. Biomass-derived syngas fermentation into biofuels: Opportunities and challenges. *Bioresource Trchnology*, 2010, 101, 5013-5022.

[99] Mohammadi M. et al. Bioconversion of synthesis gas to second generation biofuels: A review. *Renewable and Sustainable Energy Reviews*, 2011, 15, 4255- 4273.

[100] Grady J., Chen G. Bioconversion of waste biomass to useful products. US Patent No. 5821111 A, 1998.

[101] Henstra A. et al. Microbiology of synthesis gas fermentation for biofuel production. *Current Opinion in Biotechnology*, 2007, 18, 200-206.

[102] Datar R. et al. Fermentation of biomass-generated producer gas to ethanol. *Biotechnology and Bioengineering*, 2004, 86, 587-594.

[103] Ahmed A., Lewis R. Fermentation of biomass-generated synthesis gas: Effects of nitric oxide. *Biotechnology and Bioengineering*, 2006, 97, 1080-1086.

[104] Turn S.Q. et al. Fuel characteristics of processed, high-fiber sugarcane. *Fuel Processing Technology*, 2003, 81, 35-55.

[105] Najafpour G. et al. Effect of organic substrate on hydrogen production from synthesis gas using *Rhodospirillum rubrum*, in batch culture. *Biochemical Engineering Journal*, 2004, 21, 123-130.

[106] Liu K. et al. Continuous syngas fermentation for the production of ethanol, n-propanol and n-butanol. *Bioresource Technology*, 2014, 151, 69-77.

[107] Liu K. et al. Mixed culture syngas fermentation and conversion of carboxylic acids into alcohols. *Bioresource Technology*, 2014, 152, 337-346.

[108] Najafpour G., Younesi H. Ethanol and acetate synthesis from waste gas using batch culture of *Clostridium ljungdahlii*. *Enzyme and Microbial Technology*, 2006, 38, 223-228.

[109] Sharma M. et al. A critical revisit of the key parameters used to describe microbial electrochemical systems. *Electrochimica Acta*, 2014, 140, 191-208.

[110] Rabaey K. et al. Microbial electrosynthesis – Revisiting the electrical route for microbial production. *Nature Reviews Microbiology*, 2010, 8, 706-716.

[111] Li Y. et al. Metals as electron acceptors in single-chamber microbial fuel cells. *Journal of Power Sources*, 2014, 269, 430-439.

[112] Butti S. et al. Microbial electrochemical technologies with the perspective of harnessing bioenergy: Maneuvering towards upscaling. *Renewable and Sustainable Energy Reviews*, 2016, 53, 462-476.

[113] Sobieszuk P., Zamojska-Jaroszewicz A. Harvesting energy and hydrogen from microbes. *Chemical and Process Engineering*, 2012, 33, 603-610.

[114] Kracke F. et al. Microbial electron transport and energy conservation – The foundation for optimizing bioelectrochemical systems. *Frontiers in Microbiology*, 2015, 6, 575.

[115] Hernández-Fernández F. et al. Recent progress and perspectives in microbial fuel cells for bioenergy generation and wastewater treatment. *Fuel Processing Technology*, 2015, 138, 284-297.

[116] Lefebre O. et al. Microbial fuel cells for energy self-sufficient domestic wastewater treatment – A review and discussion from energetic consideration. *Applied Microbiology and Biotechnology*, 2011, 89, 259-270.

[117] Rahimnejad M. et al. Microbial fuel cell as new technology for bioelectricity generation: A review. *Alexandria Engineering Journal*, 2015, 54, 745-756.

[118] Huggins T. et al. Energy and performance comparison of microbial fuel cell and conventional aeration treating of wastewater. *Journal of Microbial and Biochemical Technology*, 2013, S6.

[119] Wang H., Ren Z. A comprehensive review of microbial electrochemical systems as a platform technology. *Biotechnology Advances*, 2013, 31, 1796-1807.

[120] Morris J. et al. Microbial fuel cell in enhancing anaerobic biodegradation of diesel. *Chemical Engineering Journal*, 2009, 146,161-167.

[121] Hu H. et al. Hydrogen production using single-chamber membrane-free microbial electrolysis cells. *Water Research*, 2008, 42, 4172-4178.

[122] Cusick R. et al. Electrochemical struvite precipitation from digestate with a fluidized bed cathode microbial electrolysis cell. *Water Research*, 2014, 54, 297-306.

[123] Rozendal R.A. et al. Efficient hydrogen peroxide generation from organic matter in a bioelectrochemical system. *Electrochemistry Communications*, 2009, 11, 1752-1755.

[124] Nevin K. et al. Microbial electrosynthesis: Feeding microbes electricity to convert carbon dioxide and water to multicarbon extracellular organic compounds. *mBio*, 2010, 1, e00103-10.

[125] Marshall C. et al. Long-term operation of microbial electrosynthesis systems improves acetate production by autotrophic microbiomes. *Environmental Science and Technology*, 2013, 47, 6023-6029.

[126] Steinbusch K. et al. Bioelectrochemical ethanol production through mediated acetate reduction by mixed cultures. *Environmental Science and Technology*, 2010, 44, 513-517.